WEEDS

a Golden Guide® from St. Martin's Press

by
ALEXANDER C. MARTIN

Illustrated by
JEAN ZALLINGER

St. Martin's Press ⚞ New York

FOREWORD

Century after century, weeds have prospered at man's expense, using the favorable habitats furnished with every expansion of agriculture. On this continent, the felling of forests and plowing of prairies paved the way for hordes of immigrant weeds that have invaded our fields, roadsides, gardens, and lawns. Today these plant pests take an annual toll from each of us, directly or indirectly.

Until recently, no one bothered about the name of the plants he destroyed. They were simply "weeds." That was all that really mattered. Now that herbicides have largely replaced the hoe and other mechanical means of control, recognizing the different kinds of weeds has become essential. Why? Because, to get desired results, specific chemicals must be used on particular kinds of weeds. This book aims to help meet the new need by acquainting readers with the most important of our unsightly, unwanted, or despised botanical intruders.

A.C.M.

ACKNOWLEDGMENTS

Substantial contributions to this book were made by: William A. Harvey, Univ. of Calif.; V. F. Bruns and Warren C. Shaw, USDA; Neil Hotchkiss, Robert D. Balcom, and Maurice N. Langley, USDI; R. R. Kriner, Rutgers Univ.; H. M. LeBaron and Earl G. Hartzfeld, Geigy Agr. Chemicals; Paul E. Warner, Phillips Petroleum Co.; and James M. Brown, Nat. Cotton Council of America; George S. Fichter; and Frank D. Venning.

CONTENTS

WHAT ARE WEEDS?

Most of the plants called weeds look "weedy." They are generally unglamorous in appearance because of their tiny flowers and unattractive foliage. A few, such as Morning-glories and Blackeyed-Susans, do have showy blossoms.

The usual definition of a weed as a "plant out of place" reflects human bias. Actually, pest plants are out of place only in terms of man's purposes. In nature's scheme, they often serve useful functions, and judging by their success in wide-open competition with other plants, they are anything but out of place.

An important factor contributing to the widespread abundance of weeds is their adaptability to diverse and adverse circumstances. Horseweeds that sometimes grow ten feet tall in favorable environment can also succeed in hard, dry soil, where plants only a few inches high may produce a crop of seeds. Other weed species can thrive even in crevices of concrete. What a contrast to our pampered cultivated plants!

Another secret of the success of weeds is their effective means of reproduction. Many of them bear tremendous quantities of seeds. Some species of Pigweed

commonly yield 100,000 or more seeds per plant. Furthermore, some weeds have remarkable devices for assuring widespread dispersal of their seeds. They either float in the slightest breeze by means of tiny, feathery parachutes or attach to clothing, fur, or wool by hooks, burs, or cleavers. Seeds of still others can remain dormant for many years until favorable conditions return.

In addition, some species can multiply by vegetative means as well as by seeds. Purslane, common in gardens and lawns, not only produces many seeds but also can start new plants readily from broken-off bits of stems that have dried for a week or two. Bermuda Grass rootstocks that have been cut off and dried can still regenerate. Chopping up the plant often simply multiplies it.

Weed seeds are remarkably adapted to assure widespread seed dispersal. Examples of special devices are shown below.

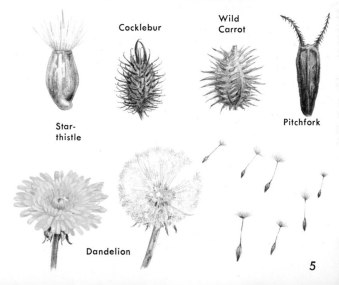

Star-thistle

Cocklebur

Wild Carrot

Pitchfork

Dandelion

THE HARM WEEDS CAUSE

Weeds cost the American farmer an estimated 5 billion dollars annually. This considers both the loss in crop yield and the expense of control measures. As consumers of agricultural products, all of us share in the paying of this huge bill.

Though this loss is tremendous, it represents only a fraction of the total harm done by pest plants. Not included in this bill, for example, is the staggering amount of time, effort, and money spent in controlling lawn and garden weeds. Nor does the agricultural loss account for the discomfort, lost time, and expense experienced by millions who suffer from hay fever and other respiratory ailments, either caused or aggravated by weed pollens. Other costs that should be added include the injury or poisoning of people and livestock by thorny or toxic weeds and the unsightliness and fire hazard created by weeds along roadsides or in vacant lots. In addition, aquatic weeds choke waterways, interfere with

A typical operation for spraying weed killers

Dredging to remove water hyacinths that clog waterway

navigation, conflict with fish, wildlife, and recreational interests, and impede malaria control. The annual cost of controlling these aquatic nuisances alone runs into the millions of dollars.

Because of the great harmfulness of weeds, many states have strict laws requiring landowners to eliminate pest species that can spread their menace elsewhere. With the population explosion making food production critically important, federal, state, and local representatives now meet in annual regional conferences to pool information and to lay plans for coordinated attacks on the common foe: weeds. Many helpful bulletins on control measures useful on particular pests are available from state experiment stations and agricultural colleges.

BENEFITS FROM WEEDS

In conflict with our purposes, weeds are harmful, but in nature, they have considerable value. One example of their value is the benefit contributed by pioneer plants that appear immediately after land is laid bare by fire, flood, axe, or plow. These plants provide a protective covering over the ground that reduces erosion. Over the seasons, they also add substantially to the organic content of the soil.

Among other values of weeds, the leaves of Wintercress, Pokeweed, and others are frequently eaten as greens. The tender underground shoots of Cattails are also used as food. Many kinds of weeds are sources of drugs, medicines, and dyes. Songbirds, gamebirds, and other kinds of wildlife depend to a very large extent on weed seeds for their existence.

Weeds have usefulness, too, as indicators of soil quality. Vigorous growths of Ragweed and Mayweed are usually found on fertile soil suitable for cultivated crops, whereas Sandbur and Povertygrass denote sterile or poor soil. An abundant stand of Sheepsorrel generally points to an acid condition that needs correcting with an application of lime.

MAJOR WEED HABITATS

Weeds can be grouped into three major habitat classifications: fields and roadsides, lawns and gardens, and marsh and aquatic areas. These divisions are arbitrary, as only a few plants are restricted totally to such habitats.

FIELD AND ROADSIDE WEEDS comprise more than two-thirds of the pest plants of the United States. These weeds are especially important economically because they affect the nation's breadbasket. In this group too are most of the weeds that produce hay-fever pollen and cause unsightly fire hazards. Because the great majority of weeds illustrated and described in this book belong in the Field and Roadside category, no separate listing of them is included here.

MARSH AND AQUATIC WEEDS occupy a habitat that varies from seasonally dry, as in temporarily flooded fields, or moist to continuously wet, as in marshes or ponds. Some of those listed below, such as Smartweeds, Barnyard Grass, Cockleburs, and Pitchforks, are also common on farmlands: Alligatorweed, Barnyard Grass, Bulrushes, Buttercups, Cattails, Cockleburs, Giant Cutgrass, Docks, Nutgrass, Pickerel-weed, Pitchforks, Reed, Rushes, Smartweeds, Spatterdock, Waterchestnut, Water Crawfoot, Waterhyacinth.

LAWN AND GARDEN WEEDS thrive in man-made, well-watered, fertile sites across the nation. Many of these weeds grow close to the ground, thus escaping mowers and sometimes even the hoe. Often they are spread by the planting of lawn grass seed, soil, or fertilizer that is contaminated with weed seeds.

Lawn and Garden weeds have wide distribution in practically every state. A few of the following also occur in farmlands: Bermudagrass, Bristlegrass, Carpetweed, Cheeses, Chickweed (Common), Chickweed (Mouseear), Crabgrass, Dandelion, English Daisy, Galinsoga, Goosegrass, Hopclover, Knotwood, Medick, Oxalis, Pepperweed, Plantains, Purslane, Quackgrass, Scarlet Pimpernel, Shepherdspurse, Speedwell, Spurges.

PLAN OF THE BOOK

This book is intended for a wide range of readers—young and mature, country and city folk, farmers, gardeners and lawn-keepers, agronomists, nature lovers, hunters, fishers, and hay-fever victims. In short, it is for the use of anyone interested in or bothered in any way by weeds. And who isn't?

PURPOSE The book's primary purpose is to aid in weed identification, and thereby weed control. To this end, full-color pictures and supplementary text are provided for 131 genera and their species, including the most common and important kinds of weeds, together with some of lesser consequence that are conspicuous regionally or locally.

MAPS Distribution within the United States of the weed genera and some of their species is indicated approximately by range maps, on which the absence of color (white) denotes absence of the weed; blue, presence to some extent; and purple, relative abundance. In most cases, the range shown is for the weed genus, though occasionally it is for a species.

ARRANGEMENT The sequence in treatment is based upon plant-family order, with the genus the usual base of discussion, as on the facing page about Cattails, genus *Typha*. In comments adjoining this range map, four species of Cattails are mentioned—but since they are restricted largely to wetland habitat, this group is limited to one page, whereas for comparatively widespread and abundant weeds such as Crabgrass, *Digitaria* (pages 22-23), Docks, *Rumex* (pages 38-39), Smartweeds, *Polygonum* (pages 40-41), and numerous other especially important pest plants, two-page coverage is given.

Typha latifolia Typha angustifolia

CATTAILS are often nuisances along irrigation ditches, in rice fields, reservoirs, and elsewhere. Even in duck marshes they may be objectionable because the dense stands prevent the growth of plants that supply food and cover for waterfowl and other kinds of wildlife. Cattail marshes are desirable for muskrat production, however, because these animals feed on the edible rootstocks. The leaves are valued, too, for matting, thatching, and other practical purposes. About a dozen species occur throughout the world; four are found in the U.S.

Broadleaf or Common Cattail, *Typha latifolia,* is widespread over the U.S. Narrowleaf Cattail, *T. angustifolia,* and *T. glauca* are in the Northeast, and *T. domingensis* in the South.

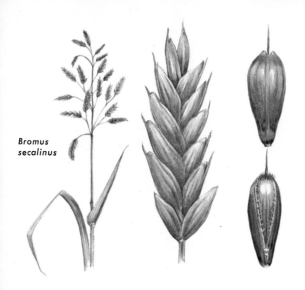

Bromus
secalinus

CHESS, also known as Brome, numbers upwards of 100 species. About a dozen in the U.S. deserve designation as weeds. These grasses are particularly plentiful in Pacific and Rocky Mountain states, where their growth over slopes and valleys provides valuable forage that compensates for their nuisance in cultivated areas. The weedy species are mainly annuals from Europe. One of the most familiar, called Cheat, is such a common pest in winter rye and wheat fields in the U.S. and Canada that many farmers believe some wheat kernels change into this weed.

Besides Cheat, *Bromus secalinus,* of grainfields, weedy species common in the West include Red Chess, *B. rubens,* Soft Chess, *B. mollis,* and Barren Chess, *B. sterilis.*

12

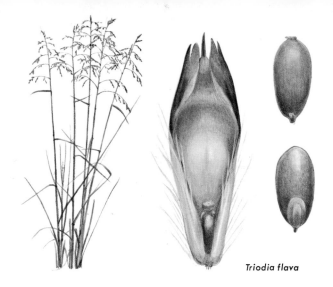

Triodia flava

TALL REDTOP, also called False Redtop or Purpletop, is a familiar feature of the rural South, where in late summer and fall much of the countryside is blanketed with its dark red-purple panicles. In northern states it occurs more sparingly. A tall (usually 2-4 feet) native perennial, it produces several stems that, at maturity, are somewhat sticky, ill-smelling, and unpalatable to livestock. These features help assure the plant's success as a weed. Except for its grace and for its attractive color in fallow fields, pastures, and roadsides, Tall Redtop is a definite liability.

Tall Redtop, *Triodia flava*, belongs to a genus of about 30 species widely distributed in temperate regions. Of several species native to the U.S., only Tall Redtop is a serious weed.

13

Phragmites communis

REED, also called Cane, Feathergrass, Carrizo (in the Southwest), and other names, is a single worldwide species. Though not as well known, it is nearly as common as the Cattails. This giant grass covers many thousands of acres of poorly drained land. It grows to 12 feet or more tall, and spreads by underground rootstocks and by leafy runners, which sometimes extend 10 feet or more from the parent plant in one growing season. In summer and fall it is topped by a large grayish plume that usually produces no seeds. The plant has minor value for matting and thatching.

Reed, *Phragmites communis,* is a liability rather than an asset for wildlife purposes. It is a pest along irrigation ditches and becomes a fire hazard. It will not thrive on well-drained soils.

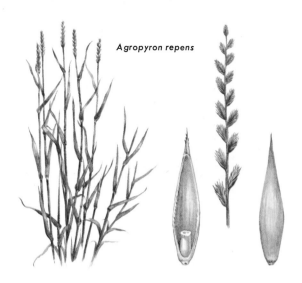

Agropyron repens

QUACKGRASS was first noted as a serious weed in the U.S. in 1837. It has since become well established in the Northeast and is locally plentiful in the prairie region and along the Pacific Coast. Besides being a nuisance in lawns and gardens, this pest plant invades vacant lots, fields, roadsides and many other places. It usually grows 1 to 2 feet tall, the seed head resembling slender wheat. Almost every piece of the tough, straw-colored rootstock can reproduce the plant. In addition to the introduced Common Quackgrass, a similar native species grows in the West.

Common Quackgrass, *Agropyron repens*, the widespread introduced species, has numerous other names, such as Couchgrass, Wheatgrass, Knotgrass, and Scotchgrass.

15

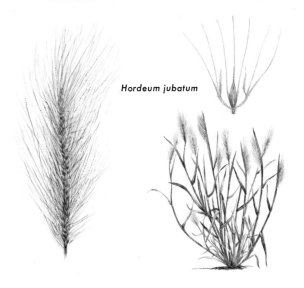

Hordeum jubatum

WILD BARLEYS, like their cultivated cousins, have a bearded seed head. The exotic species that are common in the West have such a bushy beard that these grasses are often called Squirreltail, Foxtail, Flickertail, or Skunktail. The barbed awns of their beards can be painfully harmful to livestock and are a frequent nuisance by clinging to clothing. Sheep can be blinded or choked from them. Of the twenty-odd species of Wild Barley in temperate regions, four are regarded as serious pasture weeds in the U.S. They are annuals or perennials, generally a foot or two tall.

Squirreltail species are *Hordeum jubatum* and *H. murinum*; Little Wild Barley, *H. pusillum*, favors saline soil in the West; Meadow Wild Barley, *H. nodosum*, occurs in West and Midwest.

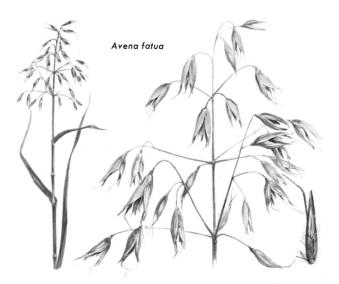

Avena fatua

WILD OATS (or Wild Oat) is one of nature's especially successful products. A native of the Old World, this aggressive annual grass is now widely established on most continents. In North America, it is particularly plentiful in Pacific Coast states and western Canada. In California, the plant's growth along roadsides, in fields, and elsewhere often rivals the density and vigor of cultivated grain crops. Wild Oats generally grows 2-3 feet tall. In especially favorable conditions, the graceful panicles of long-awned spikelets may hang 6 feet above the ground.

Seeds of Wild Oats, *Avena fatua,* are eaten freely by wildlife. A slender and less common wild species, *A. barbata,* occurs locally in Pacific states. Cultivated Oats is *A. sativa.*

17

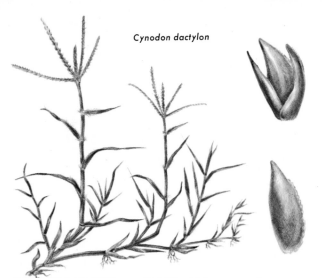

Cynodon dactylon

BERMUDA GRASS, an Old World low-growing perennial, is widely established in fields, gardens, and lawns throughout warmer parts of the U.S. In many lawns it is a serious pest, but it is also used extensively as a lawn plant in the South. Bermuda Grass loses much of its green during the winter months. It cannot survive where freezing is too severe. This grass is spread not only by seeds and surface runners but also by underground scaly rootstocks, some of which are shallow, others deep. Any part of the rootstock can grow independently. Bermuda Grass tolerates flooding.

Bermuda Grass, *Cynodon dactylon,* is valued for grazing and for stabilizing ditch banks, as well as for lawns. On alluvial soils it may grow rank enough to be cut for hay. It is the only species of *Cynodon* in the U.S.

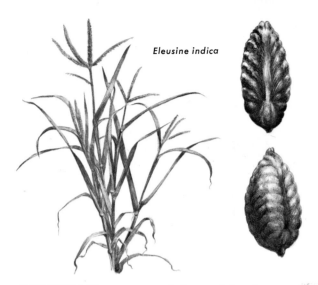

Eleusine indica

GOOSEGRASS, an annual from Asia, is our only common weedy representative of a genus of six species native to warmer parts of the Old World. It is mainly an urban weed of gardens, lawns, vacant lots, and other waste places, confined largely to the South and the Pacific Coast. Goosegrass thrives in good soils and also in packed ground, such as in paths and in some poor lawns. The elongate finger-like spikes, usually 3 to 6 in a windmill arrangement, and the flattish ascending stems are somewhat distinctive. Crowfoot Grass is another name for this plant.

Goosegrass, *Eleusine indica,* is partial to moist, rich soil. Its seeds are shaped like tiny brownish-red grains of wheat. They are marked by diagonal grooves. The seeds are commonly eaten by songbirds.

19

Zizaniopsis miliacea

GIANT CUTGRASS or Water Millet, a tall (sometimes to 13 feet), aggressive weed, covers thousands of acres of marshland in the South. It is especially abundant in old rice fields in South Carolina and Georgia, but it ranges as far northward as Maryland, Kentucky, southeastern Missouri, and Oklahoma. Added to its other liabilities is the fact that the saw-edged leaves can inflict painful cuts, making travel hazardous through dense stands. Besides growing on land it also thrives in shallow water. It spreads from stout creeping rootstalks as well as by seed.

Giant Cutgrass, *Zizaniopsis miliacea*, is also called Whitemarsh. This perennial true grass, in the family Gramineae, is native of Tropical America. Once established, it is difficult to control.

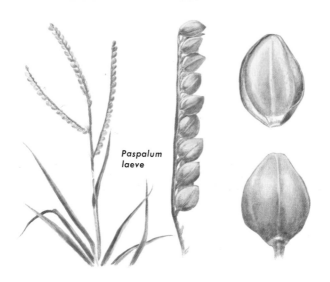

Paspalum laeve

PASPALUM, a warm-climate genus of approximately 200 species, is abundantly represented in fields and lawns throughout the South. One species is grown for forage; another is used in soil-erosion control. The few weedy species of fields and lawns include at least two that have been introduced from South America. Their distinctive circular to oval seeds (spikelets) are flat on one side and rounded or convex on the other. They are borne in rows on two, three, or more branches (racemes). Paspalums are perennials or annuals that vary in height from a few inches to four or more feet.

Dallis Grass, *Paspalum dilatatum,* and Vasey-Grass, *P. urvillei,* both from South America, are common weeds, as are the native *P. distichum,* (Knotgrass), *P. laeve,* and *P. ciliatifolium.*

Digitaria sanguinalis

CRABGRASS belongs in a genus of about 60 species of tropical and temperate regions. Only two, Small Crabgrass and Large Crabgrass, are serious weeds throughout the nation. Both are annuals from Europe, and both thrive in moist rich soils of lawns, gardens, fields, ditch banks, and roadsides. They sprout from seeds in late spring or early summer and continue growing and flowering until killed by frost in the fall, creating brown or bare patches in lawns.

Small Crabgrass, usually only a couple of inches to a half-foot high, is the worst lawn pest of the two. Setting the lawn mower high is desirable because close cutting keeps this plant from being overshadowed by other grasses. Many people confuse this plant with Bermuda Grass but the latter is perennial.

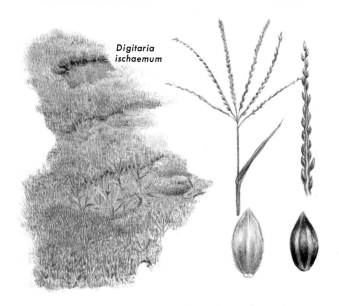

Digitaria ischaemum

Large Crabgrass, often 1 to 3 feet tall, also grows in lawns but is common in fields, gardens, and roadsides as well. Because it develops in fields after harvest, years ago it acquired the name Cropgrass. A related species is used for pasture in the tropics.

Small Crabgrass is *Digitaria ischaemum*; Large Crabgrass, *D. sanguinalis*. Low-growing native species are *D. filiformis*, *D. serotina*, and *D. laeviglumis*. Seeds of Small Crabgrass are black; those of Large Crabgrass are straw-colored and longer. The seeds of both are eaten freely by songbirds and gamebirds. Large Crabgrass is palatable for grazing and for hay.

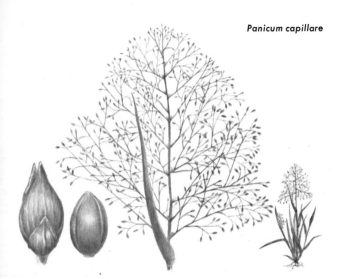

PANIC GRASS belongs to the large genus *Panicum*, with about 400 species in warm regions of the world. In the U.S., 160 species have been listed. They are most abundant in the Southeast. Fortunately, none are serious pests, and only three or four are classed as weeds of fields and gardens. Panic Grass seeds are generally borne on panicles, which in some species are intricately branched. The genus includes both annuals and perennials. Some are only a few inches tall, others several feet. Two species, Para Grass and Guinea Grass, are grown in the tropics for hay and forage.

Witchgrass, *Panicum capillare*, is an especially abundant species; a close second is Fall Panic Grass, *P. dichotomiflorum*. Switchgrass, *P. virgatum*, is a tall, stooling perennial.

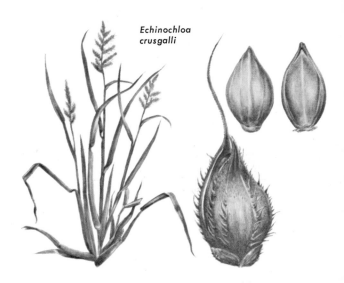

Echinochloa crusgalli

BARNYARD GRASS may be ranked either undesirable or desirable, depending on where it is growing and on the point of view. In rice fields in the South and in California, this Old World annual is a scourge to growers, yet it is a valuable asset to waterfowl and to duck hunters. Rice farmers call it Water Grass. Wildlife managers usually refer to it as Wild Millet, and they sow it in duck marshes. As its name implies, it does occur in barnyards. Barnyard Grass also grows in other upland locations but is partial to sites that are seasonally flooded.

Barnyard Grass is *Echinochloa crusgalli*. Similar closely related species are *E. walteri*, *E. colonum*, *E. frumentacea*, and *E. pungens*, with many varieties adapted to specific habitats.

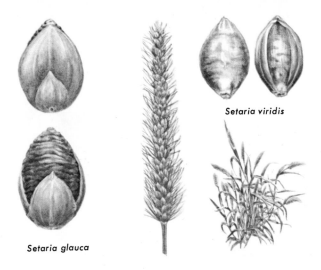

Setaria viridis

Setaria glauca

BRISTLEGRASS seed heads are like miniature bristle-brushes. Of the many species, only five are listed as weeds, but they—especially two—are so widespread and abundant that practically everyone sees them again and again. They are serious pests of croplands, gardens, meadows, lawns, and waste places, but their seeds are a valuable food for many wild birds. Bristlegrasses that are weedy are mainly European annuals a foot or two high. Nodding Foxtail grows four feet or more tall; another is perennial. Their short production season makes them undesirable pasture grasses.

The European Green Bristlegrass, *Setaria viridis,* and Yellow Bristlegrass, *S. glauca,* are especially abundant. Bur Bristlegrass, *S. verticillata,* is common. Nodding Foxtail, *S. magna,* thrives in moist meadows.

26

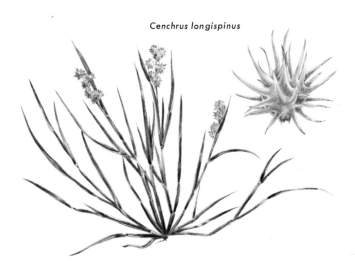

Cenchrus longispinus

SANDBUR, also known as Sandspur, gets its name from partiality to sandy soil and from its sharply armed seeds. The spiny bur is a remarkable protective feature of the approximately 25 species in this genus, which is limited mainly to tropical America. Besides being a menace to hands, feet, or other parts, Sandbur is a weedy nuisance on farms, in vacant lots, and along roadsides in various parts of the nation. It usually is a low annual, only a few inches high, with radiating prostrate or ascending branches. Livestock do not seem to like the plant as food even when it is young.

The common Sandbur of U.S. coastal areas is *Cenchrus tribuloides*. The species that occurs inland is *C. longispinus*, shown above. *C. incertus*, to 3 feet tall, inhabits sandy areas of the Southeast.

Andropogon virginicus

BROOMSEDGE, also called Broomsage, is neither a true sedge nor a sage. The "broom" part of the name is appropriate, however, for in the South, handfuls of the plants have often been used as brooms. Only the three abundant Eastern species are considered weeds; more than 30 species grow in the U.S. Broomsedges are tall (usually 2-4 feet), somewhat bushy perennials that tend to produce solid stands in fallow fields, often coloring the countryside reddish-brown. If grazed heavily, they provide useful pasture; otherwise, they become difficult-to-control pests.

The most abundant representative in the South is Virginia Broomsedge, *Andropogon virginicus*. In the East and Midwest, *A. scoparius* and *A. gerardi* are also common weeds.

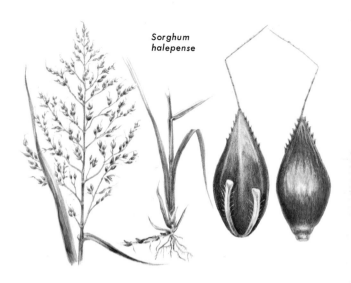

Sorghum
halepense

JOHNSON GRASS is a dubious memorial to Mr. William Johnson who, in 1840, introduced it into the U.S. from the Mediterranean area. Formerly, this tall (usually 3-5 feet) perennial was planted extensively for pasture and hay, but now, because of its persistent rootstocks, sometimes two or more feet deep, and very durable seeds, it is one of the worst weeds in the South and Southwest. Some infested farms and orchards have had to be abandoned. Furthermore, in early stages of growth and during drought, Johnson Grass can be poisonous to livestock.

Johnson Grass, *Sorghum halepense,* is a member of the sorghum group so valuable in agriculture. Herbicides strong enough to kill this weed are likely to injure nearby crops.

29

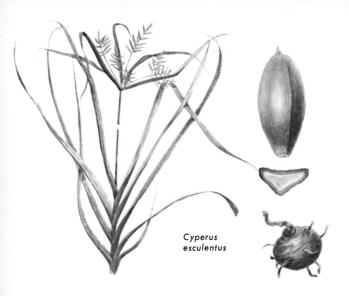

Cyperus esculentus

NUTGRASS, a member of the sedge family, Cyperaceae, belongs to a genus of about 600 species; approximately 50 occur in the U.S. Nutgrasses are those few species in the genus that possess hard underground tubers, or nutlets. These tubers can reproduce the plant when the top is cut off. This makes the plants serious pests to farmers, gardners, and lawnowners. Nutgrasses are partial to fertile, loose, or sandy soils. They are particularly prevalent in the South, especially the Mississippi Valley, and in irrigated lands of California.

Yellow Nutgrass, (also called Chufa), *Cyperus esculentus,* is the best known species and is a valued food for waterfowl. Another troublesome species is Cocosedge, *C. rotundus.*

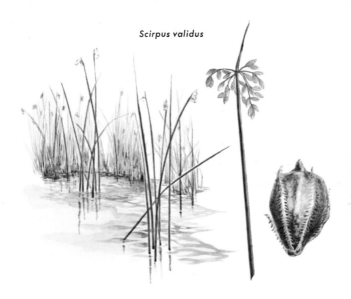

Scirpus validus

BULRUSHES total about 150 species, nearly 50 of which occur in the U.S. There are two principal types: those called Bulrush (or Tule in the West) have cylindrical bullwhip-like stems, whereas the other group, sometimes called Threesquares, has triangular stalks. Bulrushes are diverse in height, varying from a few inches or about a foot tall to nearly 10 feet. Most are not harmful, but some are nuisances in meadows, pastures, rice fields, ditches, reservoirs, and other moist areas; a few thrive in mildly saline conditions.

Common or Hardstem Bulrush is *Scirpus acutus.* Other common and important species are Woolgrass, *S. cyperinus,* Meadow Bulrush, *S. atrovirens,* and Swordgrass, *S. americanus.*

31

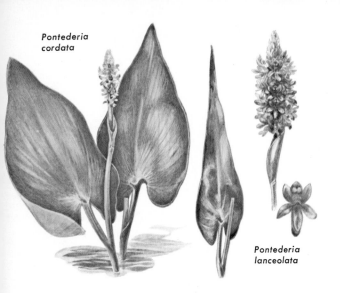

Pontederia cordata

Pontederia lanceolata

PICKERELWEED is a coarse marsh-aquatic plant limited to the East. The plant provides practically no wildlife food, clogs up streams, hampers travel of small boats, and contributes to mosquito nuisance. The most extensive beds are in fresh or mildly brackish water along the Atlantic and Gulf coasts and in the Upper Mississippi Valley. Pickerelweed's heart-shaped leaves have long petioles, raising them a few feet above water level. They lack prominent veins. The plants produce a spike of attractive blue flowers.

Common Pickerelweed, *Pontederia cordata,* is supplanted in South Florida and along the Gulf Coast by *P. lanceolata,* which also occurs locally northward to Delaware.

Eichornia crassipes

WATER HYACINTH is a tropical invader that infests the Gulf Coast region as well as many other parts of the world. It probably ranks as the worst aquatic weed in the U.S. because it hinders navigation and often clogs the passages completely. Millions of dollars have been spent in efforts to control this pest. Losses it causes to wildlife in Louisiana alone are estimated at more than 14 million dollars annually. The reproductive potential is tremendous; three plants may increase into 3,000 within 50 days.

Water Hyacinth, *Eichornia crassipes,* has minute seeds which have practically no value to wildlife. Cold weather prevents the plant from extending its range northward.

33

Juncus tenuis

RUSHES have cylindrical, pointed stems, like some of the Bulrushes. This genus of about 200 species in the family Juncaceae has numerous representatives in the U.S. Most have unbranched, hollow or pithy stems. They grow mainly in moist poorly drained places, such as meadows, pastures, and along ditches. Black Rush is common in saline coastal marshes of the Northeast, and Needle Rush dominates thousands of acres of tidal marshland from Maryland to Texas. Rushes have minute seeds. The plants have little economic value.

Black Rush, *Juncus gerardi*, Soft Rush, *J. effusus*, Baltic Rush, *J. balticus*, Yard Rush, *J. tenuis*, Toad Rush, *J. bufonius*, and Needle Rush, *J. roemerianus*, are among the important species.

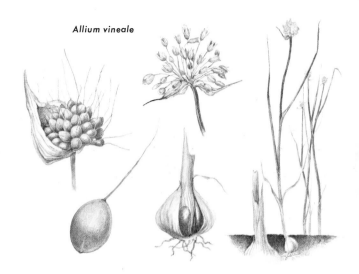

Allium vineale

WILD ONION, Wild Garlic, and their aromatic cultivated cousins comprise about 300 species in the amaryllis family. Even those who like onion or garlic flavor with meats and vegetables do not appreciate these flavors in milk or butter, and this happens when wild onions become too plentiful in pastures. In infested fields of grain, the seeds or bulblets of these plants can ruin the value of flour. Wild onions are perennials that vary from a few inches to a couple of feet high. Several are native, but the worst is a European species.

Wild Garlic, *Allium vineale*, an exotic pest, is abundant in the Southeast but is found elsewhere too. Other troublesome species are *A. canadense*, *A. tricoccum*, and *A. mutabile*.

35

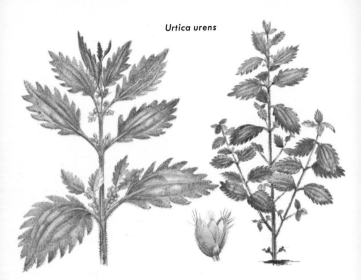

Urtica urens

STINGING NETTLES are widespread across North America but are partial to rich soil; some prefer rich damp loams. Two naturalized Eurasian species and four native ones are most common as weeds; these occur in neglected yards (often beside buildings), in waste places, along roadsides, and in bottomlands. At maturity they vary from 4 inches up to 2-12 feet tall. This enormous variation depends partly on the species, but more on the growing conditions. All have stinging hairs on the stems and leaves which sting, redden, and blister the skin if touched.

Urtica urens and U. dioica are the naturalized Eurasian species; U. gracilis, U. procera, U. lyallii, and U. holosericea are native. All but U. urens are perennials; spread by seeds, rootstocks.

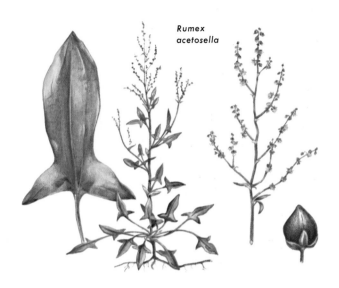

Rumex acetosella

SHEEP SORREL belongs to a genus of about 125 species native to the Northern Hemisphere. This Eurasian perennial is locally abundant throughout the U.S. in gardens, lawns, fields, and waste places. It prefers acid soil, but will thrive on neutral or slightly alkaline soils low in nitrogen. In some sites the distinctive arrow-shaped leaves assume a reddish tint. During early growth stages the plants are only a few inches high. Later, the flower stalk becomes a foot or more tall. Sheep Sorrel spreads by slender running rootstocks and by triangular brownish seeds.

Sheep Sorrel, *Rumex acetosella,* is in the buckwheat family. *Polygonaceae.* Applying nitrogen and liming acid soil controls it. Wild Sorrel, *R. hastatulus,* is taller, common in South.

Rumex crispus

DOCKS of 125 or more species occur throughout the world; about two dozen are found in the U.S. They belong to the same genus as Sheep Sorrel (p. 37) and are members of the buckwheat family, Polygonaceae. Docks are generally one to several feet tall, their oblong or lance-shaped leaves 2-6 inches long. Many small, bracted, or scaly flowers occur on upper branches. The flower clusters become brownish or reddish-brown, depending on the species. Some are native, others introduced; most of them are perennials. They are widespread, occurring in meadows, pastures, fields, along roads, occasionally in lawns, and some in swampy or marshy places. Several species are grown for their edible leaves; a few are grown as ornamentals.

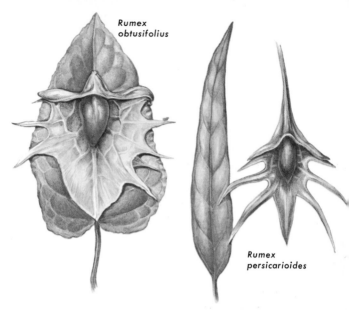

*Rumex
obtusifolius*

*Rumex
persicarioides*

Curled or Yellow Dock, widespread and familiar, is a European weed that stands 1-4 feet tall. It has wavy-curled leaf margins and a deep taproot. During first year, plant has flattish rosette of leaves. Often grows on well-drained sites.

Curled Dock is *Rumex crispus.* The common Golden Dock, *R. persicarioides,* generally found in poorly drained soil. Swamp Dock, *R. verticillatus,* grows in low, wet ground. Others are Bitter Dock, *R. obtusifolius,* Fiddle Dock, *R. pulcher,* and Tall Dock, *R. altissimus.* The Eurasian *R. acetosa* and *R. patientia,* are grown for greens.

39

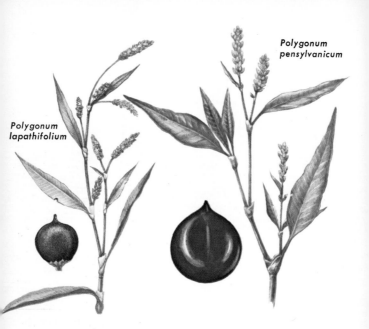

Polygonum pensylvanicum

Polygonum lapathifolium

SMARTWEEDS, members of the buckwheat family, Polygonaceae, are liabilities as weeds but valuable as sources of food for wildlife. They contain an acrid juice which causes smarting, hence the name "Smart-weed." Of about 200 species in the world, approximately two dozen are widespread in the U.S. They are partial to moist soil in pastures, cultivated fields, and along ditches. A few species thrive in water.

Smartweeds vary from a few inches to four or more feet tall. Their leaves are usually lance-shaped, and their pink, rose-colored, or white flowers are ordinarily arranged in either tight or loose spikes at the tops of the branches. Most are annuals; a few are perennials. Several Asian species are cultivated as ornamentals.

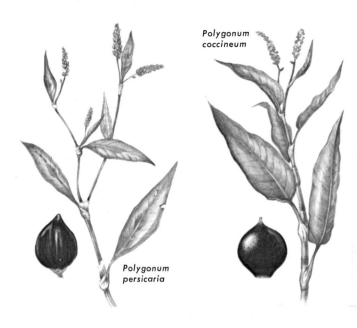

Polygonum coccineum

Polygonum persicaria

Smartweed seeds are fairly large. They vary from black to brown and from triangular to flattish. Upland gamebirds and songbirds feed on seeds; those of marshland species are eaten by waterfowl.

Common species include Big-seeded Smartweed, *Polygonum pensylvanicum*, Lady's Thumb, *P. persicaria*, and Mild Smartweed, *P. hydropiperoides*, all of which are typical of fields or moist soil. Nodding Smartweed, *P. lapathifolium* and Dotted Smartweed, *P. punctatum*, grow typically in wetlands. *P. coccineum* and *P. muhlenbergii* commonly grow as true aquatics.

41

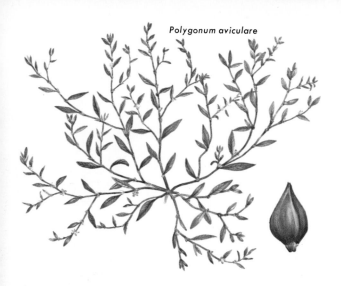

Polygonum aviculare

KNOTWEED, also known as Knotgrass, Birdgrass, Doorweed, and Waygrass, is of uncertain origin; it may have come from Eurasia. It is now widespread and locally abundant in the U.S.

This annual belongs in a genus of about 200 species. Ordinarily, the plant grows prostrate, its wiry branches and small oval leaves forming dense mats. It thrives in firmly packed soils, such as in paths or roadways. This habit is reflected in the plant's common names of Waygrass and Doorweed, for Knotweed also is a common pest in gardens and lawns.

Common Knotweed is *Polygonum aviculare*. Among the several other species in the U.S. are Erect Knotweed, *P. erectum*, and Bushy Knotweed, *P. ramosissimum*. All belong to the buckwheat family, Polygonaceae.

42

Salsola kali
var. tenuifolia

RUSSIAN THISTLE though quite spiny when mature, is not a true thistle at all; it belongs to the goosefoot family. It is from Russia, however, where vast areas of fertile farmland have been ruined by this weed. In 1873 or 1874, it was introduced accidentally into South Dakota in a shipment of flax seed. Within about 20 years, it had become established in 16 states and 13 Canadian provinces. On almost any tour of the West, the plant's "tumbleweed" debris can be seen in ditches or caught in fences. This bushy annual grows 1-3 ft. tall.

Our single species of Russian Thistle, *Salsola kali* var *tenuifolia*, is aplenty. For control plants must be destroyed before they mature; they depend on prolific seed production for spread and survival.

Chenopodium album

GOOSEFOOT is a genus of major importance, total-ing about a score of species in the U.S. These unattrac-tive plants are so prolific and prevalent that they can be seen almost anywhere during summer months—along roadsides, in fields, gardens, and waste places.

Most of the pest species are foreign annuals. They range in height from 1 to 6 feet. Their leaves are variously shaped, some showing a slight resemblance to goose feet. Some species are strongly scented, some clammy to the touch, and one is prickly. The tiny in-conspicuous flowers in crowded clusters yield myriads of seed, often tens of thousands per plant, which helps explain the abundance of these weeds through-out the nation. The small black or brownish seeds,

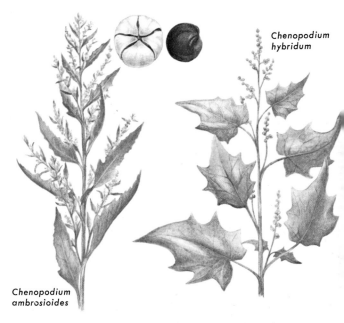

*Chenopodium
hybridum*

*Chenopodium
ambrosioides*

sometimes retained in a five-parted calyx, have a rounded or thick edge rather than narrow as in *Amaranthus*, a related genus. Controlled by hoeing or by pulling up the plants before they go to seed. Perennial Goosefoot was introduced from Europe as a pot-herb, and is still grown or gathered for greens.

Lambsquarter Goosefoot, *Chenopodium album* (leaves clammy) is widespread. Wormwood Goosefoot, *C. ambrosioides* (leaves scented) is abundant in South. Also common are Strawberry Goosefoot, *C. capitatum*; Mapleleaf Goosefoot, *C. hybridum*; Oakleaf Goosefoot, *C. glaucum*.

45

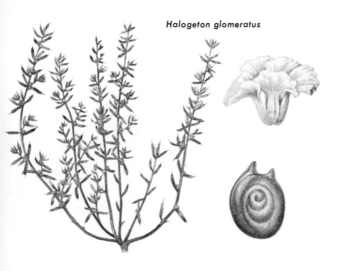

Halogeton glomeratus

HALOGETON, one of the Old World's worst weeds of saline soils, first appeared in the U.S. in 1935 at Wells, Nevada. Since then the name Halogeton has become disturbingly familiar to western ranchers and weed-control specialists, for the plant has spread widely to ten neighboring states, infesting more than two million acres of range and poisoning thousands of livestock. The plant is a low broad annual a few inches to a foot high. Each of its succulent cylindrical leaves is tipped by a single hair.

Halogeton's technical name is *Halogeton glomeratus.* Toxic oxalates occur in all of its aboveground parts. Eradication generally requires 5 to 10 years of intensive control work.

Alternanthera philoxeroides

ALLIGATOR WEED ranks next to Water Hyacinth as one of the most troublesome marsh-aquatic weeds of the Gulf and Lower Atlantic coasts. This member of the pigweed family, Amaranthaceae, is also present in Central and South America. It grows reclining on and rooted in mud, its growing tips or branches extending upward a foot or more. Broken-off joints are capable of reproducing the plant. Stems may elongate as much as 200 inches in one season. Herbicidal control is difficult because stems are partially buried under silt.

Alligator Weed, *Alternanthera philoxeroides*, a freshwater plant that tolerates mild salinity, has recently spread slightly northward. Value as waterfowl food is low.

47

Amaranthus retroflexus

Amaranthus graecizans

PIGWEED is at least as abundant in the U.S. as Goosefoot (p. 44). Again, the plant's success seems to hinge mainly on its tremendous seed production; the crop of small, shiny, black or reddish-brown seeds counted on several plants of different Pigweed species ranged from about 100,000 to nearly 200,000 on each. The seeds can remain viable but dormant in the soil for over forty years.

Pigweeds generally prefer rich cultivated soil in fields, gardens, and orchards, but some thrive in waste places and even in crevices of walks or paving. They are coarse annuals. Some species are low or prostrate; others grow 4 feet or more tall. Most are of foreign

Amaranthus spinosus

origin; a few are natives. Several species are called tumbleweeds, a name also used for other plants that break off at their base and blow about, scattering seeds until they lodge against a fence or other obstruction. Joseph's Coat, Molten Fire, Love-lies-bleeding, and Cockscomb are ornamentals in same genus.

Most prevalent species in the U.S. is Green Pigweed or Redroot, *Amaranthus retroflexus;* Tumbling Pigweed, *A. albus,* and Prostrate Pigweed, *A. graecizans,* are common in West. Spiny Pigweed (2 spines in each joint), *A. spinosus,* occurs in East.

49

Phytolacca americana

POKEWEED is so conspicuous in the South that it is known there by various names such as Poke, Pokum, Skoke, Skokum, Pokeberry, Pigeonberry, Inkberry, and Cancerjalap. Though non-woody and hollow-stemmed, it is shrubby or treelike in form, often growing about as high as a person or even higher. The roots and black clustered berries are said to be poisonous, and when robins or other birds feast on the fruits, they sometimes act as though drunk. Leaves of young Pokeweed shoots are gathered because, when well-boiled and drained, they are popular as table greens.

Only the one species, *Phytolacca americana*, grows in U.S. This native perennial can be controlled readily while young by digging. Pokeweed seldom becomes a serious weed.

50

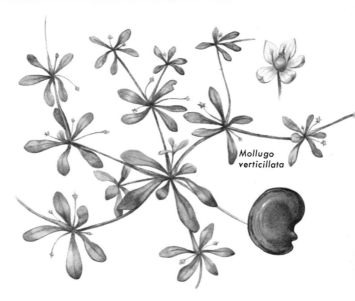

Mollugo verticillata

CARPETWEED grows in small carpet-like patches, hence its name. It also justifies the name Whorled Chickweed because its leaves, flowers, and main branches are arranged in whorls. This dainty, prostrate annual belongs to a genus of about a dozen tropical species in the carpetweed family, Aizoaceae. It is found in gardens, poor lawns, and fields throughout the U.S., seemingly favoring sandy soils. The plant has narrow spatulate leaves, about ½ inch long, and tiny white flowers from which develop oval, many-seeded capsules.

Carpetweed, *Mollugo verticillata*, produces minute, kidney-shaped, orange-brown seeds with two or three concentric ridges on outer margin. An immigrant from Tropical America.

51

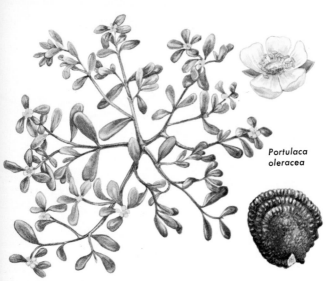

Portulaca oleracea

PURSLANE, the only weedy species among over 100 in its genus, is a pest from Europe. Most other species are harmless natives. Sometimes called Pursley, this weed is widely distributed over the world and very common in the U.S., in gardens, fields, lawns, and waste places. Though an annual and dependent on seeds for reproduction, broken-off bits of stems take root readily and start new plants. Even when hoed, the fleshy stems remain alive for a long time. The succulent wedge-shaped leaves are rounded at the tip, and the small yellow flowers stay open in bright light.

Seeds of Purslane, *Portulaca oleracea*, are flattish, kidney-shaped, and dark, with starlike bumps. Rose Moss, *P. grandiflora*, is an ornamental with red, yellow, or white flowers.

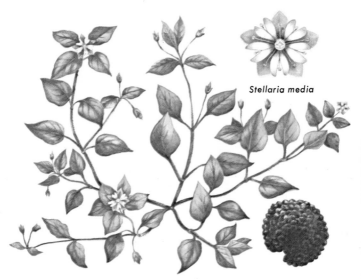

Stellaria media

COMMON CHICKWEED probably originated in Europe but is widespread in various parts of the world, including the U.S. This is the only species of about 100 in the genus that is abundant enough to be ranked as a weed. It prefers rich, moist, shaded soil in lawns and gardens, often starting its growth in late fall as a winter annual. Seeds are needed for reproduction, but the slender stems, often a foot or more long, help spread the plant by rooting at joints. Common Chickweed has opposite oval leaves and small, starlike white flowers. It is variable in size.

Common Chickweed, *Stellaria media*, has tiny, dark brown seeds covered by many small irregular bumps. Another species, Stitchwort, *S. graminea,* is locally abundant in the East.

53

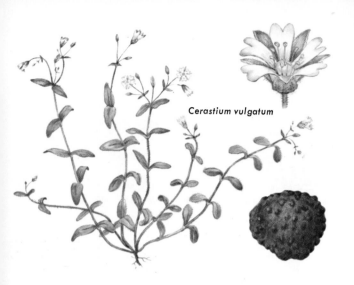

Cerastium vulgatum

MOUSE-EAR CHICKWEED is closely related to Common Chickweed but belongs to a different genus. Three species, of about 50 kinds in temperate regions, are common weeds in the U.S. as well as in other parts of the world. They usually grow in moist rich soil. As weeds, they occur mainly in gardens and lawns. Like Common Chickweed, these plants have lance-shaped or oval, opposite leaves, reclining stems that frequently root at the joints, 5-petaled small white flowers, and capsules bearing many seeds. They differ in being more hairy.

Meadow Mouse-ear Chickweed is *Cerastium arvense;* Annual Mouse-ear Chickweed, *C. viscosum;* Perennial Mouse-ear Chickweed, *C. vulgatum.* They belong to pink family, *Caryophyllaceae.*

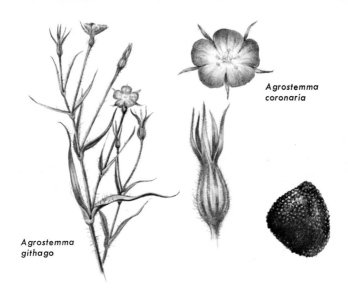

Agrostemma coronaria

Agrostemma githago

COCKLE, a tall (2-3 feet) member of the pink (carnation) family, has attractive lavender-pink flowers. It can be found in many grainfields of the East and Midwest and is especially common in winter wheat. Year after year, this Eurasian annual is re-seeded into farmlands by sowing grain that is imperfectly screened. More effective threshing, together with pulling up of the weeds before they produce their crop of seeds, will control Cockle. The large, black, sculptured seeds sometimes spoil the flavor of bread and, when numerous in screenings, are toxic to poultry.

Only the one species, *Agrostemma githago,* is widespread as a common weed in the U.S. The closely related Rose Campion, *A. coronaria,* is commonly grown in gardens.

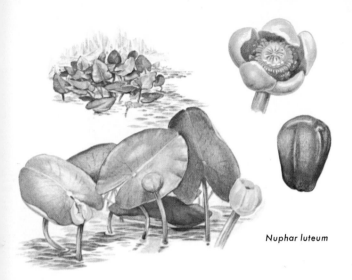

Nuphar luteum

SPATTERDOCK, also called Yellow Pond Lily and Cow Lily, is widespread in the U.S. except in Southwest. These perennials are particularly abundant in ponds, sluggish streams, canals, irrigation reservoirs, and swamps along the eastern seaboard and in the Great Lakes area. Spatterdocks have creeping cylindrical rootstocks. The leaves have a deep notch in the blade; both erect, floating, and submerged leaves are usually present. The yellow flowers, sometimes tinged with purple, usually stand above the water. Dense growth of Spatterdocks often interferes with navigation.

Spatterdock, *Nuphar luteum*, has limited value as food for wildlife and often competes with more useful plants. Its seeds are eaten only sparingly by ducks and other waterfowl.

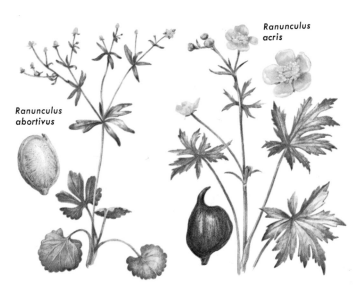

Ranunculus acris

Ranunculus abortivus

BUTTERCUPS of approximately 50 kinds are widespread in the U.S.; about 250 species are known, all from cold or temperate regions of the world. They occur in varied habitats, generally showing preference for such moist places as meadows, ditch banks, shore margins, and fresh-water marshes. Many of the more common kinds are introduced, but several are native. Some are perennial, others annual. Some are regarded as mildly poisonous. Water Crowfoots, in the same genus, are aquatics that grow largely immersed, their small, white blossoms floating on the surface.

Among the common species are Meadow Buttercup, *Ranunculus acris;* Cursed Buttercup, *R. sceleratus;* Kidneyleaf Buttercup, *R. abortivus;* Bulb Buttercup, *R. bulbosus.*

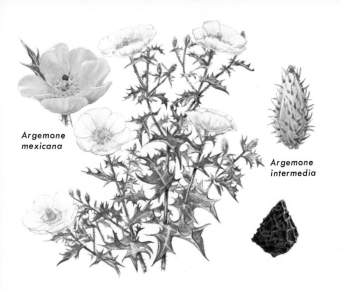

Argemone mexicana

Argemone intermedia

PRICKLY POPPY, also known as Mexican Poppy, belongs to a genus that includes about a dozen species in the American tropics and subtropics. In the southern parts of the U.S., the large white or yellow blossoms of these coarse 1- to 3-foot annuals are conspicuous in pastures, fields, and waste places. The prickly leaves are often blotched with white. Prickly Poppies are poisonous. Fortunately, livestock will ordinarily not feed on these weeds because of the painful thorns. Control depends on preventing new crops of the numerous, globose, net-sculptured seeds.

The common white-flowered Prickly Poppy is *Argemone intermedia.* The yellow-flowered *A. mexicana* occurs near the southern borders of the U.S. where it has escaped cultivation.

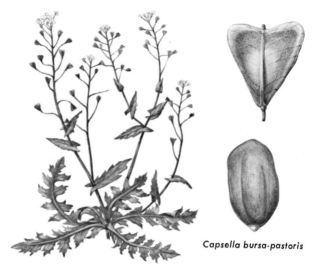

Capsella bursa-pastoris

SHEPHERD'S-PURSE, one of the most familiar of all weeds, can be found almost everywhere in the U.S. This European immigrant starts as a seedling in fall, winter, or early spring. Generally it completes its life cycle by early summer. The plants vary from a few inches to a foot or more in height, growing from a basal rosette of toothed, lobed, or variously cleft leaves. The small whitish flowers are borne in racemes that later produce distinctive wedge-shaped flattish capsules that have their broadest end at the top, suggesting an old-fashioned shepherd's purse.

Shepherd's-purse, *Capsella bursa-pastoris,* is in the mustard family, Cruciferae. Common to roadsides, lawns, and cultivated ground, it is extremely variable in foliage and shape of capsule.

59

Lepidium virginicum

Lepidium campestre

PEPPERWEED, or Peppergrass, is a genus of about 130 species in the mustard family, Cruciferae. The dozen-or-more weedy species in the U.S. are largely from Europe and are annuals or biennials. A few native species are weeds, including at least one perennial. The plants commonly have a rosette of basal leaves from which arises an erect stem, a few inches to two or more feet high. The small flowers are often crowded in clusters, with minute whitish petals, or sometimes none. The leaves vary on different species; the seed capsules are flattish and often slightly winged. The leaves and capsules have a biting peppery taste. Seeds of Pepperweeds expand and become covered with a sticky substance when moistened.

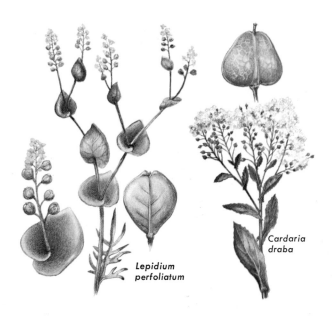

Lepidium
perfoliatum

Cardaria
draba

Pepperweeds grow in various habitats but seem partial to new lawns, meadows, and fields. Whitetop, or Hoary Cress (formerly in this genus but now classed as *Cardaria draba*) is a pest in the West.

Field Pepperweed, *Lepidium campestre,* grows in the Northeast. Greenflowered Pepperweed, *L. densiflorum,* is widespread over the U.S. Claspingleaf Pepperweed, *L. perfoliatum,* is a serious weed in the Great Basin area. Bird Pepperweed, *L. virginicum,* is also widespread. Perennial Pepperweed, *L. latifolium,* occurs in Northeast and Northwest.

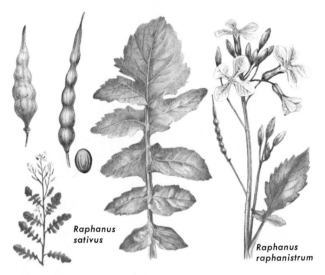

Raphanus sativus

Raphanus raphanistrum

WILD RADISH is the wild strain of plant from which, by patient selection, the garden Radish was developed. They share the same scientific name. The cultivated Radish is never a real weed. The Wild Radish, a native of Europe, is a well-established nuisance now in fields and waste places of the Pacific States, the Northeast, and some Canadian provinces. A related species, Jointed Wild Radish or Jointed Charlock, is also common locally. Its pod is constricted around each seed and breaks up into separate units. Both kinds are annuals that grow 2 to 5 feet tall.

Both Wild Radish and garden Radish are *Raphanus sativus*; Jointed Charlock is *R. raphanistrum*. In orchards along the Pacific Coast, Wild Radish is sometimes planted as a cover crop.

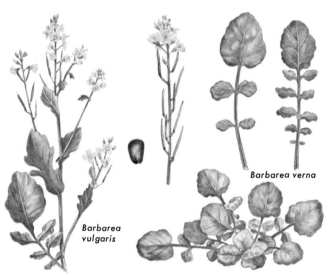

Barbarea verna

Barbarea vulgaris

WINTER CRESS includes two European species that are common now in the Northeast and Pacific region and also occur locally across the northern tier of states. Both are perennials or biennials, have bright yellow flowers, and grow a foot or two tall. In late fall and winter, the basal rosettes and young stalks can be found in meadows or in moist, rich, cultivated soil of fields and gardens. In this early stage they are often used for greens. Their main means of reproduction is by seeds, so an effective control requires destruction of the plants before they mature.

Common Winter Cress, *Barbarea vulgaris,* has 1-4 pairs of lobes on lower leaves. A double-flowered form is cultivated. The other species, *B. verna,* has 4-10 lobes on leaves.

63

Brassica rapa

MUSTARD is a name shared by the great plant family known technically as Cruciferae, by the mustard genus of about 80 species, and by the popular seasoning extracted from the plant's seeds. Mustard oil, another derivative, is used in medicines and in soap. The leaves of at least one species have been used by American Indians and others for greens. Obviously, this weed genus has considerable in its favor.

All kinds of mustard are yellow-flowered, and all have four narrow-based petals typical of the family. Our half-dozen or so species originated abroad and are mainly bushy annuals, usually varying from 1-6 feet tall. Some are found in practically every section of the

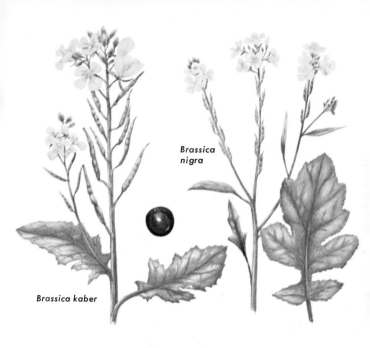

Brassica nigra

Brassica kaber

U.S., but the genus is especially conspicuous in the Northeast and in California, where it covers large areas of the countryside with beautiful yellow in spring and summer, partly due to planting of mustard as a cover crop in some orchards. Nevertheless, these weeds cause serious losses to grain and flax crops. The myriad of tiny spherical seeds, like miniature shot, account primarily for the abundance of the plants.

Black Mustard, *Brassica nigra*, one of the most widespread species, is the chief source of mustard. Other weedy species are *B. kaber* var *pinnatifida*, *B. rapa*, *B. incana*, and *B. hirta*.

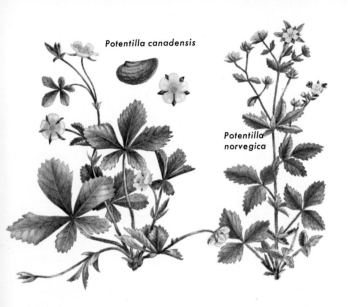

Potentilla canadensis

Potentilla norvegica

CINQUEFOIL is a temperate-region genus of about 300 species with over 33 representatives in the U.S., particularly in the Northeast and Lake States. Because many are five-leafleted, they are often called Five-finger. This name does not apply well to species with three to seven leaflets, of course. Cinquefoils that are nuisances in pastures, fields, or lawns are mostly of foreign origin; a few native species are weedy. The species vary in habit. Most are prostrate with short vinelike branches; some are erect, a foot or more tall.

Silvery Cinquefoil, *Potentilla argentea*, and Canada Cinquefoil, *P. canadensis*, are especially common. Rough Cinquefoil, *P. norvegica*, is an erect, shrubby type, to 3 feet tall.

Trifolium procumbens

HOP-CLOVERS belong to the clover genus of about 300 species, but they are distinct in having yellow flowers. The three species in the U.S. are all annuals and came from Europe. Two are widespread over the U.S. Large Hop-clover is restricted chiefly to the Northeast. These are obscure, largely unobserved plants, but one, Little Hop-clover, is a somewhat famous contender for recognition as Irish Shamrock. Hop-clovers are low or prostrate plants found mainly in lawns, pastures, and gardens, sometimes in cultivated fields.

Large Hop-clover, *Trifolium agrarium*, is the largest species. Low Hop-clover, *T. procumbens*, has distinct flower heads; Little Hop-clover, *T. dubium*, does not. All three are spread from seed.

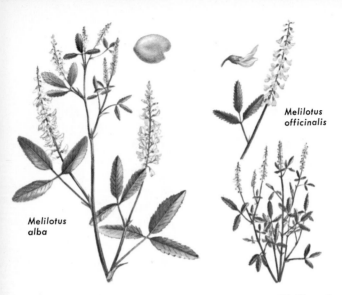

Melilotus
officinalis

Melilotus
alba

SWEET CLOVER belongs to a different genus (*Melilotus*) than do the true clovers (*Trifolium*). Of about 20 species in the world, four occur in the U.S. Two are common— one white-flowered and the other yellow. Two other yellow-flowered species are also found in the U.S. but are less common. All four are introduced biennials that grow 1- to 4-feet tall. They are abundant along roadsides and in waste places due to their widespread use for pasture, hay, or cover crops. In addition to their value as honey producers, these beneficial weeds add nitrogen to the soil by nodules on their roots.

White-flowered Sweet Clover, *Melilotus alba*, is the most commonly seen; *M. officinalis* is the prevalent yellow-flowered kind. *M. indica* and *M. altissima*, both yellow, are less common.

Medicago hispida

BUR CLOVER belongs to an Old World genus of about 65 species which includes Alfalfa, one of our most important agricultural crops. Two rather common weeds of lawns and other places are members of the same genus: Black Medick and Spotted Medick (p. 70). Bur Clover grows in practically every state but is particularly abundant in California where, following rains, its seeds often accumulate in drifts by the thousands. Plants vary from prostrate to semi-erect, generally a few inches high. Most distinctive is the spirally coiled burlike seedpod bordered by curved prickles.

Bur Clover, *Medicago hispida*, is a weed in fields, gardens, and lawns but also has value as food for livestock and wildlife. It is planted as a soil-enriching cover crop.

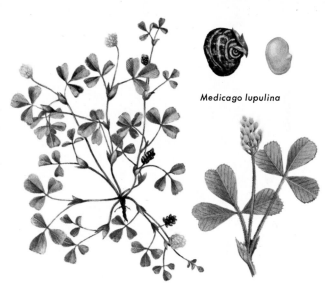

Medicago lupulina

BLACK MEDICK is an inconspicuous, low, clover-like plant. Though widespread over the U.S. and common in lawns, roadsides, and fields, it is a little-known weed, seldom seen by the casual observer. Like the approximately 65 other species in this genus (including Alfalfa and Bur Clover), Black Medick originated in Eurasia. To its credit, it has value as forage and has been used as a cover crop. The 2-3 inch, prostrate branches bear 3-foliate leaves and clusters of small yellow flowers that produce black, 1-seeded pods.

Black Medick is *Medicago lupulina*. Spotted Medick, *M. arabica*, has a dark spot in the center of each leaf, and seeds are separated from each other by partitions in pod. Belong to pea family, *Leguminosae*.

Astragalus mollissimus

LOCOWEED is the name for several closely related legumes native to semiarid and arid regions of the West. These deep-rooted perennials, usually from 6- to 18-inches tall, contain the narcotic locoine, poisonous to horses, cattle, and sheep. Although plant is at first unpalatable, animals (especially horses) become addicted. Its cumulative effects depress and debilitate the animals, distort their vision, produce a slow staggering gait, and cause them to act crazy or "loco"; they leap high over pebbles and shy easily.

Woolly Loco, *Astragalus mollissimus*, Purple Loco, *A. diphysus*, and White Loco, *Oxytropis lambertii*, all purple-flowered, are the most notorious. Other species are purple, pale yellow, or white-flowered.

71

Desmodium canadense

BEGGARWEED is also called Beggar-lice, Beggar-ticks, Sticktights, and Tick Trefoil, each name referring to the seedpod segments that cling to socks and other clothing. This bean family *(Leguminosae)* genus includes about 160 species in various continents. About 50 grow in the U.S. Though they are abundant in old fields and along woods margins in the Southeast, these plants are usually not serious weeds. Beggarweeds vary from 1- to 4-feet tall, with trifoliate leaves and small pink, lavender, or white flowers.

Most common are *Desmodium nudiflorum, D. dillenii, D. canadense,* and *D. perplexum.* Florida Beggarweed, *D. tortuosum,* is cultivated as a forage crop but becomes a pest.

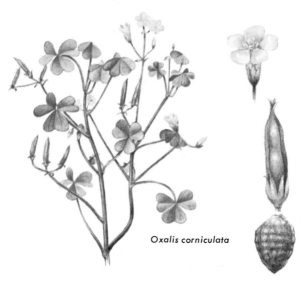

Oxalis corniculata

OXALIS, Wood Sorrel, Wood Shamrock, or Sourgrass are names given to plants in the large genus *Oxalis*, which includes about 500 species. Some are wildflowers, others grown as ornamentals for their white, pink, red, yellow, or violet flowers. Their trifoliate leaves have a pleasant acid taste. Only a few native yellow-flowered perennial species are weeds, but these are widespread. Some species spread by runners and are hard to eradicate. The brownish, flat, oval-pointed seeds are unique in having six or more cross ridges.

Erect Oxalis, *Oxalis stricta*, Lady Wood Sorrel, *O. corniculata*, along with *O. florida* and *O. europaea* (a N. Amer. native) are common weeds. *O. cernua*, locally in California.

FILAREES are beneficial weeds in the sense that, in some places, their nuisance is compensated by the value as forage to livestock and wildlife. Half a dozen species occur in the U.S., but only one is widely distributed. This species is common in the West. On the Pacific Coast, where it is particularly abundant, it constitutes much of the springtime covering in fields, orchards, gardens, lawns, open ranges, and waste places. Filarees have a central, pointed structure, the "storksbill," from which the seeds peel off spirally. Storksbill is another common name for these plants.

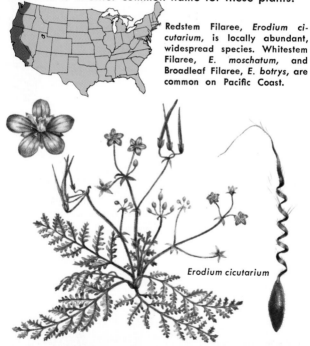

Redstem Filaree, *Erodium cicutarium*, is locally abundant, widespread species. Whitestem Filaree, *E. moschatum*, and Broadleaf Filaree, *E. botrys*, are common on Pacific Coast.

Erodium cicutarium

WILD GERANIUMS total about 260 species of temperate regions around the world. (These are not the familiar cultivated "geraniums," which belong instead in the genus *Pelargonium*.) Some are woodland flowers; only a few are weeds. Among these is Carolina Cranesbill, a native annual or biennial that occurs commonly across the U.S. in lawns, gardens, and fields. It varies in height from a few inches to nearly a foot, has attractive circular deeply cut leaves, and small pale pink flowers followed by beaklike fruits. It is spread by seeds, as are other members of the genus.

Carolina Cranesbill, *Geranium carolinianum*, and three locally common species from Europe (*G. molle, G. pusillum,* and *G. dissectum*) are among the weedy species, growing in open places. All are spread by seed.

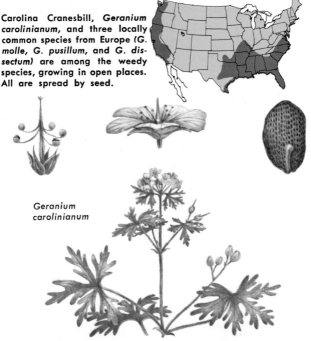

Geranium carolinianum

75

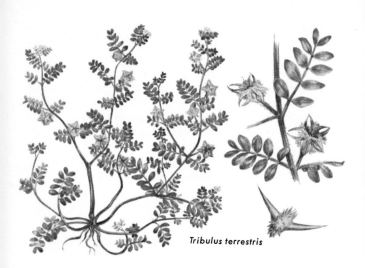

Tribulus terrestris

PUNCTURE VINE, also called Tackweed, Caltrop, and Ground Bur-nut, is an immigrant from the Mediterranean. It occurs sparsely in the East but is a serious pest in some warmer parts of the West. Usually prostrate, the plant has viny branches 1-5 feet long. When growing in crops like Alfalfa, the branches often turn upward or are erect. Two-spined nutlets are borne in groups of five and can remain alive in the ground for several years. The spines can hurt livestock, bare feet, or bicycle tires but are not a problem for auto tires. It has paired, compound leaves and yellow flowers.

Puncture Vine, *Tribulus terrestris*, is a member of the caltrop family, *Zygophyllaceae*, which includes mainly tropical herbs, shrubs, and trees. It is an annual, hence prevention of seed production is important.

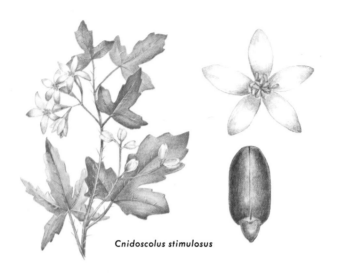

Cnidoscolus stimulosus

BULL NETTLE, also called Spurge Nettle and Tread-Softly, an herbaceous plant of the spurge family *(Euphorbiaceae),* reproduces by bug-shaped seeds and by deep, persistent roots. It prefers dry sandy soil and is a weed in cultivated fields, pastures, lawns, and gardens from Virginia and Florida westward to Texas. The plants vary from 6 inches to 3 feet tall; the upper stem may be simple or branched. The whole plant is covered with stiff, glassy, stinging hairs that break off in the skin and cannot be removed easily. The white fragrant flowers are followed by 3-seeded capsules.

Bull Nettle, *Cnidoscolus stimulosus,* is the only U.S. species of this genus of mostly tropical American plants, all similarly armed. Bull Nettle is hard to eradicate because of its deep root system.

DOVEWEEDS include about 600 species, mostly tropical. About a dozen occur in the U.S., confined largely to the South and the Southwest. The species vary in size and shape. Some are low-growing and compact; others grow tall and openly branched. Doves feed on seeds, as do quail and other wildlife. When livestock eat plants, not particularly palatable because of fuzzy coating of hairs, they are sometimes poisoned. Seeds are roundish or oval with flattish faces that fit together in a globular capsule.

The most widespread species is *Croton glandulosus.* Others are: *C. monanthogynus, C. capitatus, C. punctatus, C. lindheimeri,* and *C. texensis.* Doveweeds belong to spurge family, *Euphorbiaceae.*

COPPERLEAF is a genus of about 200 species of the tropics and subtropics. About half a dozen species are common weeds of pastures, cultivated fields, gardens, and waste places in the South. Copperleaf belongs to the spurge family, *Euphorbiaceae.* It's small oval seeds are eaten freely by gamebirds and songbirds. Copperleaf usually grows a foot or two high, with toothed lanceolate or ovate leaves that often turn copper color when mature. All of the species that grow in the U.S. are annuals. They may be very abundant locally, but are generally not serious weeds.

Common species are *Acalypha virginica, A. rhomboidea, A. gracilens,* and *A. ostryaefolia.* Because of their conspicuous winglike bracts, the plants are also called Mercury-weed.

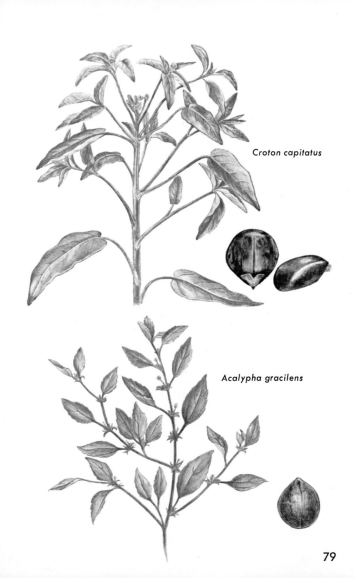

Croton capitatus

Acalypha gracilens

79

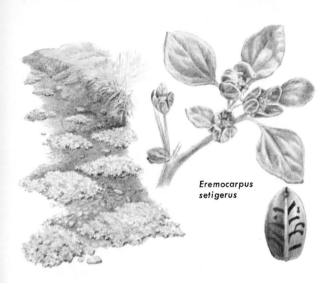

Eremocarpus setigerus

TURKEY MULLEIN, a very abundant weed on the Pacific Coast, is closely related to the Doveweeds (p. 78) and formerly was classed with them. The low, rounded, grayish-hairy clumps, generally a few inches to a foot high and a foot or two broad, are evident along roadsides and in fallow fields, from the valley floor to hillslope. The seeds are a valuable food for California Quail as well as for other wildlife, but the plant is not eaten by livestock. The stems and leaves are harsh to the touch because of their dense covering of branched bristly hairs.

Turkey Mullein, *Eremocarpus setigerus*, was formerly *Croton setigerus*. Because this weed is an annual, its oddly marked wedge-shaped seeds are its only means of distribution.

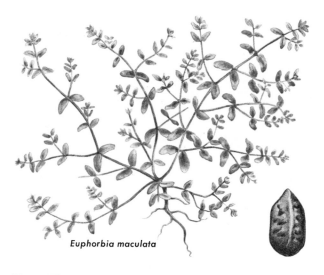

Euphorbia maculata

SPURGES are plants of diverse appearance in the genus Euphorbia, which contains about 1,000 species. About 12 are common, widespread, especially bothersome weeds. All of these are grouped in the sub-genus Chamaesyce, and are low or prostrate matted plants with acrid milky juice that may irritate the skin on contact. All have numerous clusters of tiny male and female flowers that lack both sepals and petals. The female flowers mature into small 3-celled capsules, normally with one seed in each cell. They seed so abundantly that they are difficult to eradicate.

Prostrate Spurge, *Euphorbia supina* and Spotted Spurge, *E. maculata,* are mat-formers. Erect species include Leafy, *E. esula,* Flowering, *E. corallata,* and Sun, *E. helioscopia.*

Rhus radicans

POISON IVY is clearly one of the worst of all weeds, harmful not only to the farmer but also to the general public. Anyone susceptible to the itchy rash that comes from contact with this woody weed should learn how to recognize the plant—and then avoid it. Prevention is easier than a cure. Washing thoroughly after contact with the plant is practical, and various helpful remedies are available. Jewelweed juice is said to give relief. Poison Ivy is either an upright shrub, called Poison Oak, or a high-climbing vine with aerial roots.

Poison Ivy is *Rhus radicans* (or *Toxicodendron radicans*). It involves several varieties (or species). In the East, it grows mainly as a vine; in the West, it is mainly a shrub.

Malva neglecta

MALLOWS, sometimes called Cheeses, are familiar to most of us because of their distinctive appearance and prevalence from coast to coast on farms, in gardens, and in new or poor lawns. Of about 30 Old World species in this genus, seven occur in the U.S. Introduced as ornamentals, they are usually not serious weeds. Mallows are annuals or biennials, a few inches to a foot or more tall, depending on the species, with circular, shallow-lobed, toothed leaves. Seeds are in cheese-shaped disks.

Common representatives are: *Malva rotundifolia, M. neglecta,* and *M. sylvestris.* Less common but larger are *M. nicaeensis* and *M. verticillata.* They belong to the family Malvaceae.

83

Sida spinosa

SIDA, also known as False Mallow, is a genus of about 100 species, most of which grow in the tropics or the subtropics. A dozen or more kinds occur in the U.S.; some are annuals, others perennials. Only Spiny Sida is abundant enough to be ranked as a serious weed. This species is particularly common in cultivated fields, gardens, and waste places in the South. The plant generally grows a foot or two tall, with toothed, ovate-lanceolate, alternate leaves. The small, pale-yellow, mallow-like flowers produce two-spined containers with triangular seeds.

Spiny Sida is *Sida spinosa*. Another less common species in the South is *S. rhombifolia*. *S. hederacea* is a semi-woody perennial that occurs in saline areas in the West.

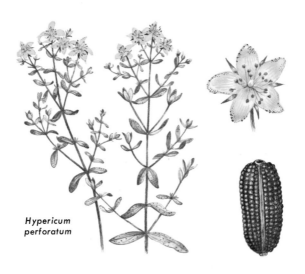

Hypericum perforatum

KLAMATHWEED is a European immigrant now widely established across the northern U.S. It is an aggressive pest in the Pacific Northwest, from Oregon to central, coastal California, invading waste places and pastures. The semi-woody perennial plants grow 1- to 5-feet tall, with oblong, finely dotted leaves and clusters of showy yellow flowers. Though not preferred by livestock, it is eaten if other forage is scarce. Some animals, when exposed to bright sunlight after eating large quantities, develop a rash, blister, and shed hair, especially around the nose, eyes, and ears.

Of 200 or more species in this genus, Klamathweed, *Hypericum perforatum*, is the only important weed. It is also very troublesome in Australia. Some species are ornamentals.

WATER CHESTNUT, a Eurasian aquatic pest, was first found in the U.S. about 1884 near Scotia, N.Y. Since that time it has spread widely throughout the Northeast and may be locally abundant. A few years ago, 2,500 acres on the Hudson and Mohawk rivers were covered by this plant's surface rosettes. Its flexible stems extend downward from the floating rosettes of leaves and small flowers through 2 to 15 of water to the roots. Once established, it is very difficult to control. The large fruits armed with very sharp spines are a menace on bathing beaches.

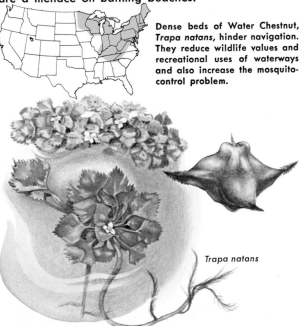

Dense beds of Water Chestnut, *Trapa natans*, hinder navigation. They reduce wildlife values and recreational uses of waterways and also increase the mosquito-control problem.

Trapa natans

Oenothera biennis

EVENING-PRIMROSE is a genus of about 200 New World species, several of which are weeds. The plants vary from annuals or biennials to perennials. Some are stemless with flowers in a basal rosette; others have aerial stems, either branched or unbranched. They have flowers that are yellow or white, with a distinctive four-branched style, all located above an elongate ovary. Field Primrose, an especially common widespread species of pastures and other fields, produces a rosette of long leaves the first year and a tall, erect, flowering stem in the second.

Field Primrose, *Oenothera biennis,* has been the subject of genetic studies. Other common species are O. *missouriensis,* O. *fruticosa,* O. *laciniata,* and O. *humifusa.*

WILD CARROT, also called Queen Anne's Lace, is a wild Eurasian variety of the common cultivated carrot. They share the same scientific name. Wild Carrot is widespread across the nation, but it is especially abundant in the Northeast, where it is sometimes profuse in fallow fields, pastures, fence rows, and waste places. A biennial, its 1-3 foot stems and lacy flowers do not appear until second year. Usually a tiny dark blossom marks center of flower head. Dried-up flower head suggests one of plant's common names: Bird's Nest.

Wild Carrot, *Daucus carota*, is most widespread species. In West, the smaller *D. pusillus* is widespread but not ordinarily a serious weed. Control depends largely on preventing seed production.

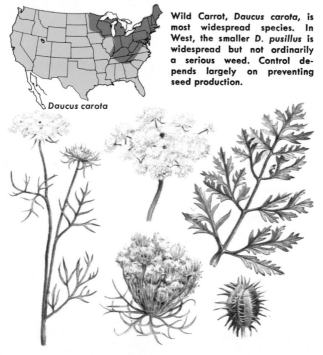

Daucus carota

88

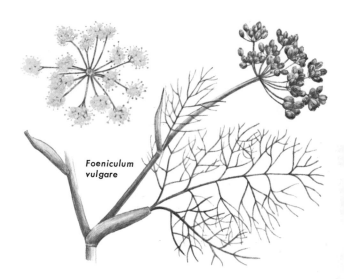

Foeniculum vulgare

SWEET FENNEL, a tall (1-6 foot) Mediterranean perennial, has spread all over the world, is found throughout the U.S., and is locally abundant in California pastures and waste places. A thin waxy coating on the plant's coarse stems and finely divided leaves gives them a bluish-green cast. All parts of the plant have a pleasant anise-like aroma and taste. Fennel is essential to modern Italian and French cookery; the seeds are used in cooking, for liqueurs, and for candy. The oil is used in soaps, medicines, and perfumes. The leaves have a distinctive, long, sheathing base.

Sweet Fennel is *Foeniculum vulgare,* and belongs to the family Umbelliferae. Unfortunately, the name "fennel" is also applied to two unrelated plants that have finely divided leaves.

89

Anagallis arvensis

SCARLET PIMPERNEL is widespread but rarely a problem plant. Though a common invader of gardens, lawns, fields, and waste places, this European annual deserves to be considered a wildflower because of its attractive brick-red blossoms that shut quickly at the approach of bad weather; hence in England it is called "Poor Man's Weather-glass." In its low or prostrate habit and small oval leaves it resembles Common Chickweed, and is known as Red Chickweed and Poison Chickweed. The many small seeds are borne in a globular capsule that splits open in the middle.

Scarlet Pimpernel, *Anagallis arvensis,* suspected of being somewhat poisonous to livestock, is the only species in its genus in the U.S. It belongs to the primrose family, Primulaceae.

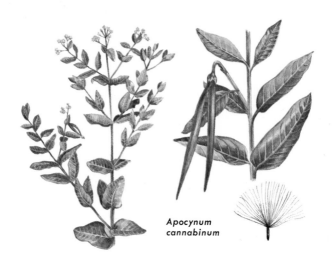

Apocynum cannabinum

DOGBANE is a reddish-stemmed native perennial, 2-4 feet tall. It has oval-oblong light green leaves and bears long dangling pods. The tough fibrous bark was used by the Indians, hence the names Indian Hemp and America Hemp have been used for this plant. The names Indian Physic and Rheumatism Weed refer to the plant's medicinal properties, but it is dangerously poisonous to cattle, horses, and sheep. Dogbane resembles milkweeds in having a milky juice and a tuft of silky hairs on the end of the rodlike seeds, but it belongs to a different family, Apocynaceae.

Common Dogbane is *Apocynum cannabinum*; there are several less abundant species. Common Dogbane grows in fallow fields and in pastures. It is not an aggressive weed.

91

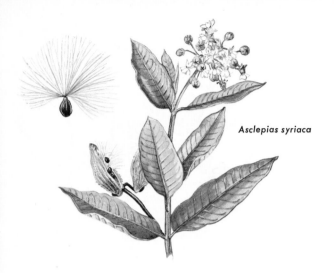

Asclepias syriaca

MILKWEED is a genus of about 150 species. Many are native to the U.S. They are the familiar, coarse, milky-juice perennials that grow in pastures, fallow fields, and along fence rows and roadsides. A few kinds inhabit marshes or other wet places. Their dense broad heads of small flowers, usually pink or reddish-purple, are replaced later by large pods filled with many flattish seeds with silky plumes. To control, plants must be killed before pods mature and seeds are blown far and wide. The beautiful orange-flowered Butterflyweed of the East belongs to this genus.

Showy Milkweed, *Asclepias speciosa*, occurs throughout much of the nation; Common Milkweed, *A. syriaca* of eastern U.S., and Narrowleaf Milkweed, *A. verticillata*, are principal species.

MORNING-GLORY, as dealt with here, comprises two distinct genera, *Ipomoea* and *Convolvulus*. Both are large groups, totally including about 600 species. Bindweed (p. 94), of greater importance, is treated separately. Morning-glories include both native and exotic species common throughout the nation. They are especially abundant in old fields and waste places in the Southeast. Some are annuals, others perennials. Generally, they are not serious weeds. Included in the group are the beautiful blue, purple, red, pink or white-flowered vines grown on trellises or fences.

Common Morning-glory, *Ipomoea purpurea*, Red Morning-glory, *I. coccinea*, Wild Potato-vine, *I. pandurata*, and Hedge Bindweed, *Convolvulus sepium*, are common species.

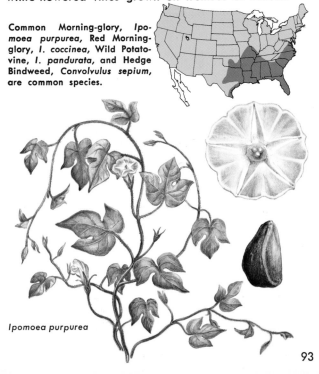

Ipomoea purpurea

BINDWEED is one of the most difficult of all weeds to eradicate. Once established, it takes years to eliminate, either with sprays or by manual or mechanical means. The reason is that the plants have a very extensive root system, sometimes penetrating to a depth of 10 feet, with lateral branches at various depths. Each lateral branch is a potential new plant. This Eurasian immigrant, now widespread in the U.S., makes mats of growth in old fields, orchards, and waste places. Bindweed belongs in the morning-glory family, *Convolvulaceae.*

Bindweed, *Convolvulus arvensis,* is also called Creeping Jenny, Field Morning-glory, or sometimes Field Bindweed to distinguish it from other plants also known as bindweeds.

Convolvulus arvensis

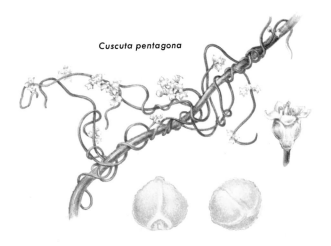

Cuscuta pentagona

DODDERS are annual, leafless, parasitic vines belonging to the cosmopolitan genus *Cuscuta* in the morning-glory family. Eight native and three introduced species are pests of herbaceous crops such as flax, alfalfa, clover, peas, and beans; some attack tree crops. Dodders lack chlorophyll; the seeds germinate in soil, but the young plants quickly twine onto the host plants, adhering by suckers. They also spread virus diseases from one host to another. The profuse clusters of tiny whitish flowers are usually produced in summer and fall.

Flax Dodder, *Cuscuta epilinum,* from Europe, and Alfalfa D., *C. suaveolens,* from S.A., are noxious weeds. Clover D., *C. epithymum,* from Europe, and the native *C. indecora* are crop pests.

GROMWELL, also called Puccoon, Stoneseed, and Wheatthief, is a small genus that includes several native species particularly common in the West. Two widespread European immigrants are clearly weeds. They occur commonly in grainfields, pastures, waste places, and along railroad tracks. The erect plants, either branched or unbranched at the base, grow 1-3 feet tall. They have lanceolate or linear leaves and small creamy or yellowish flowers clustered near the branch tips. Gromwells are annuals or perennials and can be controlled readily by tilling.

Common Gromwell, *Lithospermum arvense*, is troublesome in fields where winter rye or wheat are planted each year. Pearl Gromwell, *L. officinale*, is the other weedy European species.

FIDDLENECK gets its name from the fiddle-like arching of its inflorescence. The small yellow flowers produce somewhat triangular stony nutlets that are valuable food for gamebirds and other wildlife. Fiddlenecks occur locally in a number of parts of the U.S. They are especially abundant in Pacific Coast states, where they flourish in grainfields, orchards, vineyards, meadows, and waste places. The plants grow 1-2 feet tall, either branched or unbranched at the base. Fiddlenecks usually have lanceolate leaves with bristly hairs. They are controlled by tilling.

Douglas Fiddleneck, *Amsinckia douglasiana*, is widespread. Others: Coast Fiddleneck, *A. intermedia*, Tesselate Fiddleneck, *A. tesselata*, Bugle Fiddleneck, *A. lycopsoides*.

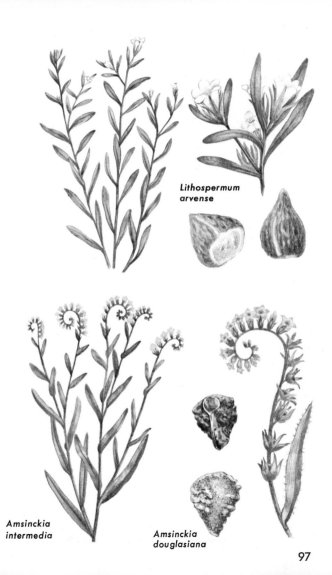

*Lithospermum
arvense*

*Amsinckia
intermedia*

*Amsinckia
douglasiana*

97

Verbena stricta

VERVAIN is a genus of 75-100 species found mainly in the Americas. Besides weedy species, some of which are attractive and properly regarded as wildflowers, the beautiful Verbenas of the flower garden belong to this genus.

Three perennial kinds of weedy Vervains are common to abundant in the East and less common farther westward. Most of the species in the South and in the Southwest are annuals. Their usual habitat consists of pastures, moist meadows, old fields, and waste places, and they do well in soils ranging from rich to gravelly.

Some species, especially the annuals, are prostrate. Others grow erect, usually 1-4 feet tall and with small bluish, purple, or white flowers in dense or loose

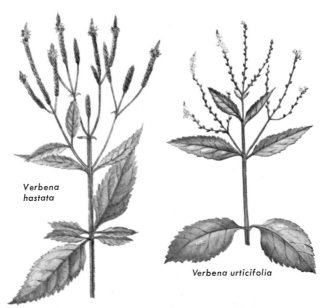

Verbena hastata

Verbena urticifolia

spikes that are produced all summer long. The leaves may be lanceolate to ovate in erect perennial forms; in annuals, they are generally dissected. The seeds are oblong, flattish nutlets, about 1/16 of an inch long. They are eaten by birds.

Blue Vervain, *Verbena hastata,* widespread in the U.S., is especially common in Northeast. Hoary Vervain, *V. stricta,* smaller with showy flowers, is abundant from Lake States westward. White Vervain, *V. urticifolia* is found along woods. Prostrate Vervain, *V. bracteata,* occurs in South and Southwest; European Vervain, *V. officinalis,* occurs in the Southeast.

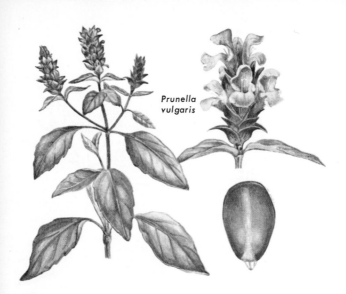

Prunella vulgaris

HEAL-ALL grows in most sections of the U.S. from Maine to Florida and California and also in many other countries. It is believed to be native to the Northern Hemisphere of both New and Old Worlds. A member of the mint family, Labiatae, it has the family's characteristic "square" or 4-angled stems and the irregular 2-lipped flowers, which in Heal-all are small, purple to pink, and borne in bracted, thick, close spikes. This erect or sometimes reclining low perennial grows in pastures, meadows, lawns, and waste places. Its distinctive seeds are marked by lengthwise lines.

Heal-all, *Prunella vulgaris*, is also called Self-heal. In moist lawns it spreads rapidly by rooting at each stem joint. Difficult to dig out, it is controlled by using herbicides.

100

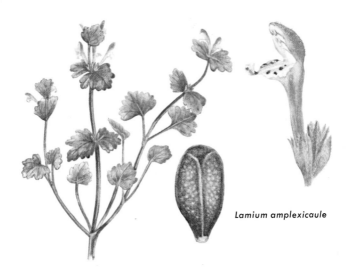

Lamium amplexicaule

HENBIT, also called Dead Nettle, is a familiar weed of rich, moist soil in gardens, cultivated fields, and waste places. The three common species in the U.S. were introduced from Europe and Asia, where they are also common plants. The plants are a few inches to a foot or two tall. They have 4-angled stems, opposite heart-shaped or broadly ovate leaves, and 2-lipped flowers in whorls at leaf axils, typical of members of the mint family. Henbits are annuals, biennials, or perennials. Though widespread, they are rarely serious pests or difficult to control.

Common Henbit, *Lamium amplexicaule*, Red Henbit, *L. purpureum*, and Spotted Henbit, *L. maculatum*, are common species. Red and Spotted names refer to flower characteristics.

HOREHOUND, a European perennial, is widespread in the U.S. and particularly abundant in California. Like many other mints, it has 4-angled stems, opposite leaves, irregular 2-lipped corollas, and a distinctive aroma. The stems are whitish, and the dark gray-green, veiny, crinkled leaves are whitish-woolly below. The tiny white flowers are in whorls in the leaf axils. The recurved teeth of the calyx become attached to hairs of animals, helping to spread seeds. Common around old buildings and on ditch banks, roadsides, fallow fields, pastures, and waste places.

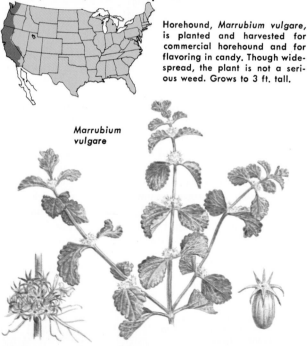

Horehound, *Marrubium vulgare*, is planted and harvested for commercial horehound and for flavoring in candy. Though widespread, the plant is not a serious weed. Grows to 3 ft. tall.

Marrubium vulgare

JIMSONWEED is also called Thorn-Apple, Jamestown Weed, Devil's Trumpet, and Stinkweed. Its crushed leaves have a rank odor. Poisonous and narcotic, the plant is a commercial source of hyoscyamine, atropine, and scopolamine. This coarse treelike annual, 1-4 feet tall, is native to Asia but occurs as a weed in yards, fields, and waste places all over the world. It has alternate, ovate, toothed leaves and large trumpet-shaped white or purplish blossoms 3 to 4 inches long. The dark brown, flat, kidney-shaped seeds are borne in a prickly oval-shaped capsule.

Jimsonweed, *Datura stramonium*, is one of 12-15 species in genus. Similar species are *D. ferox* from Asia, *D. innoxia* from Tropical America, and *D. meteloides* of southwestern U.S.

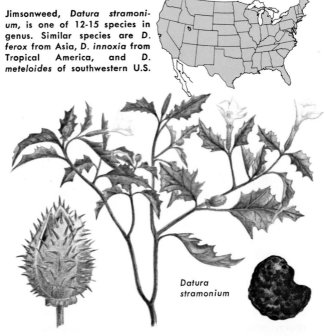

Datura stramonium

Solanum carolinense

NIGHTSHADE is the common name applied to a number of weedy species of the genus *Solanum*. Some go by such common names as Horse Nettle, Bull Nettle, Bittersweet, Buffalo Bur, and Black Nightshade. *Solanum* is a large genus of about 1,500 species and includes such important food plants as Irish Potato and Eggplant, and several beautiful ornamental vines and shrubs. All are members of the nightshade or potato family, Solanaceae.

Weedy nightshades of both native and introduced species are widespread over the U.S. They are extremely variable: some are annuals, others perennials; some grow moderately low or have a vinelike habit, others are erect and moderately tall; some species are spiny, others unarmed. All of them have alternate

104

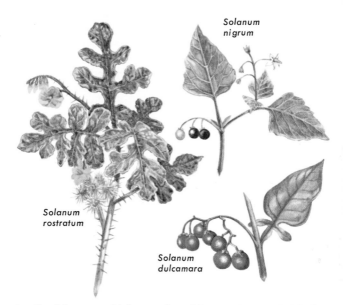

Solanum nigrum

Solanum rostratum

Solanum dulcamara

toothed leaves which may be either entire or divided. The flowers are usually white or yellow. They produce berries that may be yellow, black, or red, smooth in some species, covered with prickles in others. The spiny species are especially widespread and are the most difficult to control or eradicate.

Black Nightshade, *Solanum nigrum*, is a common exotic species. An edible, cultivated variety is called Wonderberry. Bittersweet, *S. dulcamara*, is a red-berried species that grows on moist soils in Northeast. Bull Nettle *S. elaeagnifolium*, and Buffalo Bur, *S. rostratum*, are weeds of the Southwest.

GROUNDCHERRY belongs to the genus *Physalis,* containing about 80 species that occur mainly in North America. They are widely distributed over the U.S. Many are perennial natives. A few are annuals, some exotic. Most species are erect, branched, low plants, 1-2 feet high, with small flowers and somewhat cherry-like berries of yellow, orange, or purple. Ground-cherries are common in pastures, cultivated fields, and waste places. Fruits and seeds are eaten by wildlife. Abundant locally and widespread, these plants are easily controlled and are seldom serious pests.

Groundcherry common species include *Physalis virginiana, P. heterophylla, P. pumila,* and *P. lanceolata.* Chinese Lantern, *P. alkekengi,* is an ornamental that sometimes becomes a weed.

MULLEIN is an Old World genus of about 250 species. Only two, Common Mullein and Moth Mullein, are well established as weeds in the U.S. These two look so unlike each other that they do not appear closely related, except for flower structure and the netted surface of their seeds. Common Mullein is a coarse, stout, biennial of pastures and fallow fields. The first year it produces a large rosette of feltlike hairy leaves. The following year it produces a stem 1-7 feet tall, often branched like a candelabra. Moth Mullein, also a biennial, is more slender and smooth.

Common Mullein, *Verbascum thapsus,* and Moth Mullein, *V. blattaria,* are both common throughout the U.S. They prosper in sunny spots, often growing on otherwise bare soil. Usually neither is a serious pest plant.

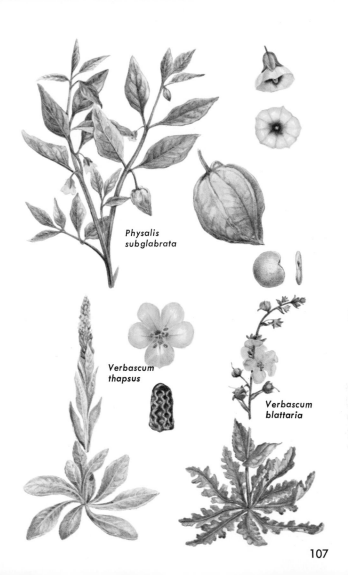

*Physalis
subglabrata*

*Verbascum
thapsus*

*Verbascum
blattaria*

107

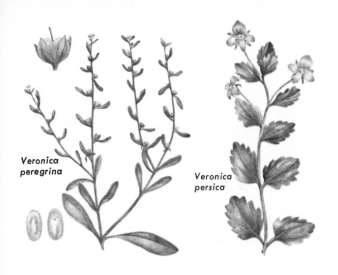

Veronica peregrina

Veronica persica

SPEEDWELL belongs to a widely distributed group of about 150 species, a dozen or more of them weeds in various parts of the U.S. These small herbs are common in gardens, lawns, and elsewhere. Some grow only an inch or two tall; others grow to six inches or taller. Variation in foliage and other characteristics is so great among the different species that, except for technical botanical features, these plants and some closely related woody ornamentals in this group do not seem to belong to the same genus. Their flowers range in color from white or light pink to blue.

Corn Speedwell, *Veronica arvensis*, Purslane Speedwell, *V. peregrina*, Gipsy Speedwell, *V. officinalis*, and Thyme Speedwell, *V. serpyllifolia*, are among the most common species.

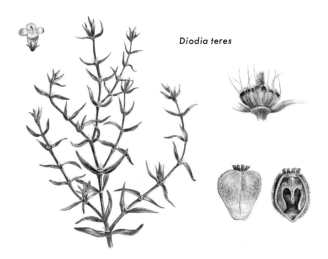

Diodia teres

BUTTONWEED, also known as Poor Joe and Poverty Weed, is partial to dry, sandy soils in cultivated fields and waste places. Buttonweed cannot tolerate the competition of larger shading plants on rich soils. This native annual is widespread in the East but is particularly abundant in the Southeastern Coastal Plain, where its low branching growth often produces a dense matting in fallow fields. The plant's linear-lanceolate stiff leaves, about an inch long, are opposite. In the leaf axils, it bears small, paired, whitish flowers that produce woody seeds.

Common Buttonweed is *Diodia teres*. A species of moist areas in the Southeast is Virginia Buttonweed, *D. virginiana*, with larger leaves and strongly furrowed fruits.

109

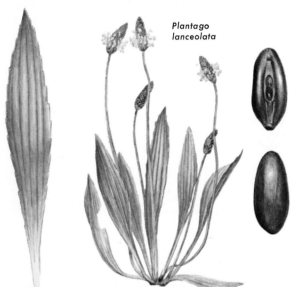

Plantago lanceolata

PLANTAINS are serious pests in lawns and are also conspicuous in other places. The genus includes perennials and annuals or biennials, some of them native and others exotic. In the typical plant, one or several erect stalks terminated by a flower spike of variable length arise from a group of basal leaves. The seed capsules split open horizontally, usually near the middle, exposing either numerous irregular seeds or a pair of somewhat boat-shaped seeds. Plantain pollen is recognized as a potent cause of hay fever. Six species of the approximately 200 in the world are prevalent in parts of the U.S.; especially common weeds are Buckhorn Plantain and Broadleaf Plantain. Both are perennials from Europe.

Buckhorn (Narrow-leafed or English) Plantain has narrow lengthwise-ribbed leaves, often a half-foot or

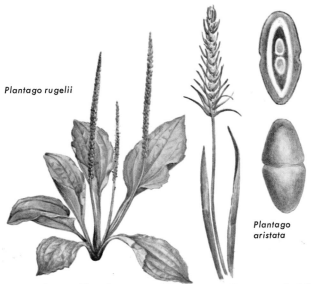

Plantago rugelii

Plantago aristata

more long. The flower stalks frequently exceed 12 inches in height. It occurs in lawns, fields, yards, pastures, and along roadsides throughout the nation.

Broadleaf (Dooryard) Plantain has broadly ovate leaves that are usually flattened toward the ground. Though it is widespread, Broadleaf Plantain appears to be most abundant in the Northeast.

Buckhorn is *Plantago lanceolata*; Broadleaf Plantain is *P. major*. A similar species, *P. rugelii*, is largely in the East. Bracted Plantain, *P. aristata*, and Hoary Plantain, *P. virginica*, and Whorled Plantain, *P. indica*, are common in the East and local elsewhere. Woolly Plantain, *P. purshii*, is largely western.

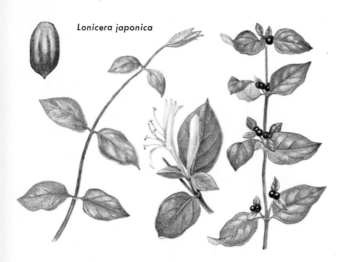

Lonicera japonica

JAPANESE HONEYSUCKLE was introduced from East Asia as an ornamental vine, prized for the fragrance of its beautiful white flowers which fade to yellow on ageing. Both gamebirds and songbirds eat the black fruits and spread the seed; the plant now occurs locally throughout the U.S. but is most abundant in the Southeast, where it is a pernicious weed. It is partial to fence rows, field borders, and woodlands; it climbs in profusion over shrubs and small trees, overwhelming and strangling the native vegetation. Yet it is still extensively planted as an ornamental.

Japanese Honeysuckle, *Lonicera japonica*, is extremely difficult to control or eradicate once established. Five more introduced species of *Lonicera* have escaped and are becoming naturalized weeds in the East.

TEASEL is a tall (2-8 feet), prickly, stiff European biennial. Its dried spiny heads were formerly used in carding wool, which earned it the name of Fuller's Teasel. Because of its unusual and attractive appearance, the plant is now widely sought for decorative purposes as a dry ornament.

Partial to rich moist soil, Teasel is locally abundant in pastures, meadows, and waste places in some states in the Northeast and also along the Pacific Coast. The basal leaves are in a rosette; those of the stem are opposite and joined cuplike at the base.

Common Teasel or Fuller's Teasel is *Dipsacus fullonum*. Other less common but similar species are Wild Teasel, *D. sylvestris*, and *D. laciniatus*. All are of European origin; none is a serious pest.

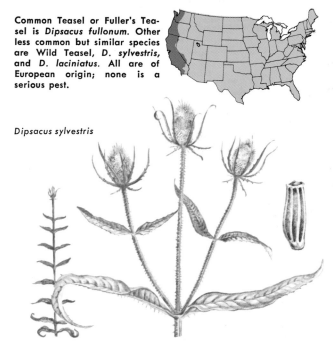

Dipsacus sylvestris

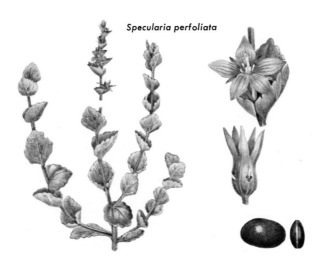

Specularia perfoliata

VENUS'S LOOKING-GLASS grows in a wide variety of habitats: rich garden soil, pastures, cultivated fields, dry gravelly sites, and waste places. Usually these annual plants are only a few inches high; occasionally they grow a foot or two tall. The plants are upright and either branched or unbranched. The most common species have alternate rounded leaves with a scalloped margin and a clasping base. Other species have ovate or linear leaves. The saucer-shaped blue or purplish flowers are borne in the axils of upper leaves. They are members of the bluebell family, Campanulaceae.

Widespread important species are *Specularia perfoliata* and *S. biflora*. Other species common in the U.S. are *S. leptocarpa* and *S. holtzingeri*. The European *S. speculum-veneris* is casual near U.S. ports.

Lobelia inflata

INDIAN TOBACCO is one of more than 200 species in the genus *Lobelia*, widely distributed in the world. A native annual, it is most abundant in the East, where it grows in meadows, pastures, cultivated fields, open woods, and gardens. The erect plants, to 2 feet tall, have alternate, ovate to oblong, toothed leaves. Small pale violet to whitish or pinkish flowers are borne near the ends of the branches. The characteristic inflated capsules with their many minute seeds develop from the flowers. The plants contain poisonous alkaloids, which have been useful in medicines.

Indian Tobacco, *Lobelia inflata*, belongs to the bluebell family, which includes numerous ornamental plants. Control is not difficult, particularly if the weed is killed before it can produce more seeds.

115

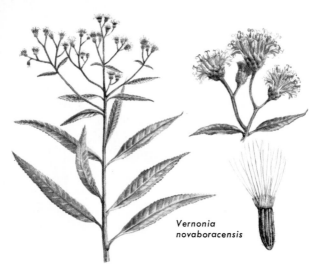

*Vernonia
novaboracensis*

IRONWEED is a tall (usually 2 to 5 feet), aggressive weed of ditch banks and other moist soils in the East and in a few states of the Midwest. Its opposite lance-shaped leaves are borne on stout stems which, in summer and fall, are topped by numerous small heads of attractive dark-purplish flowers. The plants spread by both seeds and runners. Once well established, they are not easy to eradicate. Like many other members of the daisy family, Compositae, the seeds are equipped with parachutes of bristly hairs that carry potential new plants far and wide.

Common species of Ironweed in eastern and midwestern U.S. include *Vernonia novaboracensis* (which includes a white-flowered form), *V. altissima, V. crinita V. baldwini,* and *V. missurica.* Unpalatable to livestock.

116

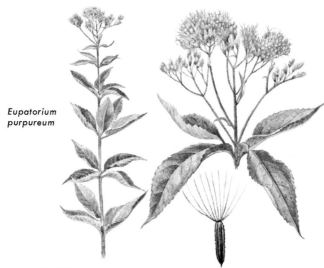

Eupatorium purpureum

JOE-PYE-WEED belongs in a genus of about 500 species, mostly native to temperate and tropical America. It is also called Purple Boneset or Tall Boneset, but these names may be confusing because other species are also called Boneset (p. 119). The plants are tall (1-5 feet), their stems greenish or speckled with dark purple. The leaves, generally lance-shaped, are arranged in whorls of 3's, 4's or 5's, and the small purple flower heads are in a flattish mass at the top of the plant. The oblong seeds are five-angled, with the usual daisy-family parachute of hairs (pappus).

The name Joe-Pye-Weed is applied to several species, including *Eupatorium purpureum*, *E. maculatum*, and *E. altissimum*. They are common in pastures and along ditch banks in the East and Midwest.

117

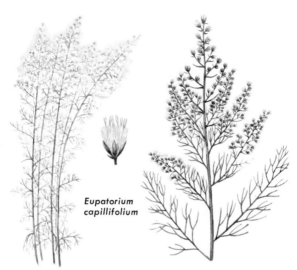

Eupatorium capillifolium

DOG FENNEL, also called Summer Cedar, is a characteristic weed of the Old South, where it is common along roadsides, in fallow fields, and waste places. If you see Dog Fennel, chances are you are near or below the Mason-Dixon line. The tall plants, usually 2 to 6 feet high, have an attractive feathery appearance due to their finely cut lacy leaves. Many small flower heads of greenish-white are scattered over the upper parts of the stems. These are followed in late summer or fall by fluffy, white masses of tiny parachute-borne seeds.

Dog Fennel, *Eupatorium capillifolium*, is the only native annual in this genus that does not persist as a weed in cultivated fields. If plowing is not practicable, the plants must be grubbed out.

118

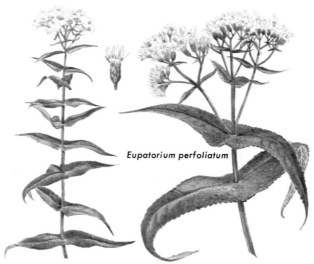

Eupatorium perfoliatum

BONESET is also called Thoroughwort, Fever-weed, Sweating Plant, and Agueweed, all suggesting medicinal properties that were formerly important. The plants are quite hairy, usually 1 to 3 feet tall. The leaves are opposite and united at their base (perfoliate) around the stem. The small white flowers in large flattish masses appear from July through October. The oblong, 4- or 5-sided seeds are carried by a parachute of bristly hairs (pappus). Boneset is partial to moist places, such as ditch banks, marsh margins, low-lying pastures, or waste places.

Boneset, *Eupatorium perfoliatum*, is one of the most common roadside plants in moist or shady places in the East. It does not persist in cultivated well-drained fields. One form has purple-tinged flowers.

119

Solidago canadensis

GOLDENRODS total at least 125 species, most of them North American. Though they are represented in practically all parts of the U.S., including marshes and deserts, their region of greatest abundance is the Northeast. Here, in particular, goldenrods may be troublesome weeds as well as attractive perennial wildflowers. In the fall, they blanket much of the countryside with brilliant yellow.

In addition to being a nuisance to farmers in their pastures, hay meadows, and other places, goldenrods are also a cause of hay fever for many. This undoubtedly helps explain why they are not grown more extensively as ornamentals.

Goldenrods vary in height, but most species grow 2-4 feet tall. They vary also in foliage, which is mainly lance-shaped, and in the arrangement of their terminal

120

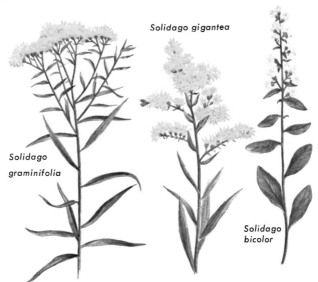

Solidago gigantea

Solidago graminifolia

Solidago bicolor

masses of small yellow flower heads. Besides reproducing by tiny seeds with hairy parachutes, they also spread by rootstocks. The plants are perennial. Although the aerial parts are killed by frost each Fall, they sprout from the roots again in spring.

Canada Goldenrod, *Solidago canadensis,* is one of the most abundant and attractive species. Common in wooded areas are Woodland Goldenrod, *S. nemoralis, S. speciosa,* and *S. perlonga;* in marshes, Seaside Goldenrod, *S. sempervirens, S. salicina,* and *S. elliottii.* White Goldenrod (Silver-rod), *S. bicolor,* has whitish flowers. Sweet Goldenrod, *S. odora,* has anise-scented foliage.

Grindelia camporum

Grindelia squarrosa

GUMWEED is also called Rosinweed or Tarweed because of sticky material on the bracts (modified leaves) of its flower heads. Mainly a western plant, it grows in the prairies, plains, and mountain regions—in pastures, abandoned fields, and along roadsides. This native perennial is usually 2 to 3 feet tall, with stiff sometimes reddish stems and alternate sharply toothed leaves. Bright yellow flowers borne individually at the ends of branches later yield oblong, somewhat 4-angled seeds, the pappus reduced to 2-8 deciduous awns. Cultivation or mowing give effective control.

Common Gumweed, *Grindelia squarrosa*, is the only one of several native species in this genus that is widely important as a weed. Another, *G. camporum*, is sometimes a weed problem in California.

122

Bellis perennis

ENGLISH DAISY is a cultivated plant that has become naturalized as a weed in lawns, golf courses, pastures, and similar places. It is common in Pacific Coast states and is locally abundant elsewhere. Some lawns become heavily infested with the plant. The blossoms are attractive, but homeowners generally prefer pure grass. English Daisy is a low perennial, sometimes to 6 inches tall if not mowed. Its oval leaves are confined to a basal rosette, from which the stout flower stalks arise. The flower heads, borne singly, measure an inch or more across.

English Daisy, *Bellis perennis*, has double or single flowers. They vary in color from white to bright pink or purplish. Some varieties have incurved, reflexed, or quill-like petals. Seeds often contaminants in grass seed.

123

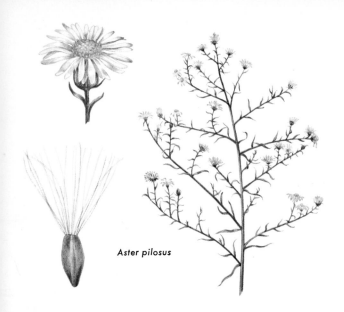

Aster pilosus

ASTERS, also called Michaelmas Daisies or Wild Daisies, include over 600 species, most of them North American. All but a few species are perennials. Asters are widespread across the U.S., along roadsides, in fields, woods, marshes, and even deserts. Like goldenrods, they are especially plentiful in the Northeast, where they are classed as weeds. Some people regard them as a source of hay fever. Many species are attractive wildflowers. Their masses of white, bluish, lavender, or pink blossoms add beauty to the fall countryside. Some 14 native and 8 introduced species are frequently cultivated as garden flowers.

Asters are usually erect and often bushy. The different species vary considerably in height. Some grow only a few inches tall, others to several feet. Their leaves

Aster laterflorus

Aster novae-angliae

range from linear or lance-shaped to ovate, and the flower heads of the different species may be tiny, medium-sized, or large and showy. Aster seeds are an important fall and winter food for songbirds.

Many-flowered or White Heath Aster, *Aster ericoides*, stores selenium from the soil. Hay containing the plant may be poisonous to livestock. Other common and similar species of open places are *A. dumosus* and *A. pilosus*. Asters of woodlands include *A. divaricatus*, *A. cordifolius*, and *A. azureus*; of marshes, *A. puniceus*, *A. junciformis*, and *A. nemoralis*; of deserts, Mexican Devilweed, *A. spinosus*, and Saline Aster, *A. exilis*.

125

FLEABANE (Whiteweed, Whitetop, or Daisy Fleabane) is a genus of about 150 species scattered over the world, mostly in temperate and mountainous regions. Some are annuals, others perennials. Though widespread in the U.S., they are particularly abundant in hayfields, pastures, and roadsides of the Northeast. Species with a few large flowerheads are attractive; 13 of them are grown as garden flowers in the U.S. Those called Whiteweed or Whitetop branch freely and have many small flowerheads. They look "weedy" but are usually pests only in hayfields.

Rough Daisy Fleabane, *Erigeron strigosus,* and *E. philadelphicus* are common weedy species. Robin's plantain, *E. pulchellus,* and Seaside Daisy, *E. glaucus,* are among the attractive cultivated native species.

HORSEWEEDS, also called Mare's-tail, Hogweed, and Butterweed, is a coarse, tall (3-6 feet), weedy-looking fleabane with many small whitish flowerheads in an extensively branched inflorescence. The seeds, with their bristly pappus, are well adapted for dispersal by the wind. A native annual occurring in practically all parts of the U.S., it is especially common in harvested fields, sometimes producing nearly solid stands. Quickly invades mismanaged or overgrazed meadows and pastures; sometimes troublesome in hayfields.

The most common Horseweed in the U.S. is *Erigeron canadensis;* in the Midwest, *E. divaricatus* is common. Control includes early mowing of infested meadows and hayfields, and clean cultivation of row crops.

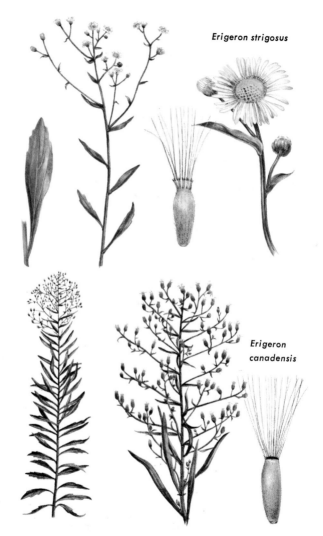

Erigeron strigosus

Erigeron canadensis

127

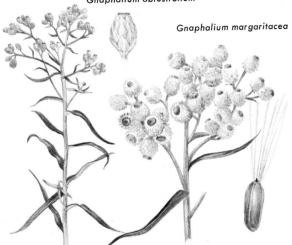

Gnaphalium obtusifolium

Gnaphalium margaritacea

CUDWEEDS, also called Everlastings or Fragrant Everlastings, include about 120 species. Some are native, others are introduced, a few of them widely distributed in the U.S. Typically, they occur in old fields, pastures, and waste places. These gray-woolly rather fragrant annuals or biennials of the daisy family (Compositae) vary from a few inches to 2 or 3 feet tall. Their roundish papery flower heads lack petals. The minute seeds have a fine, bristly, hairy parachute (pappus) that does not persist. Plants are sometimes collected as "everlasting" ornamentals.

Gnaphalium obtusifolium and *G. macounii*, common biennial species, have only a basal rosette of leaves the first year. *G. purpureum* is biennial or annual; *G. uliginosum* is a common annual garden weed.

Franseria discolor

BUR-RAGWEED, also called Burroweed or Poverty Weed, is represented by a number of species in the West, mainly in plains, Rocky Mountain, and desert areas. Eight species are found in California. They are most abundant in pastures or open ranges, making these unsatisfactory for grazing. The large quantities of pollen released causes hay fever, like true Ragweed (p. 130), which they also resemble with their dissected leaves and burlike seeds (fruits). The plants grow 1-2 feet tall, the stems are usually whitish with woolly or short hairs.

Bur-ragweed species include *Franseria discolor*, *F. dumosa*, *F. tomentosa*, and *F. acantho-carpa*. None of these species is classed as a serious weed. The closely related *F. tenuifolia* is a noxious weed in Arizona.

Ambrosia artemisiifolia

RAGWEED is a very familiar name, particularly to the millions who suffer seasonally from air-borne pollen. Yet probably few recognize the offending plant when they see it. Several species are involved, but Common Ragweed (above left) is by far the most widespread, abundant, and detrimental. It is sometimes called Hayfever Weed due to its infestation of the atmosphere by myriads of pollen grains in summer and early fall. It is also an obnoxious farm weed. This grayish-green, cut-leafed, native annual is partial to grainfields, roadsides, and waste places in the Northeast and Midwest, occurring somewhat more sparsely in other parts of the nation. Ragweed often develops solid luxuriant stands in fields after a grain crop has been harvested.

Ambrosia trifida

On the credit side, Ragweed is a boon to wildlife, for its oil-rich seeds (achenes) are a valuable source of fall and winter food for many kinds of birds. After a heavy snowfall, birds can be seen picking the seeds from branches of Ragweeds that stick above snow.

Common Ragweed, *Ambrosia artemisiifolia*, above left, and Western Ragweed, *A. psilostachya*, are the most widespread of the species. Shown at the right above is Giant Ragweed, *A. trifida*, often growing as much at 10 ft. tall. This species is partial to the rich soil of field margins in the East. *A. aptera* is common in the Southwest.

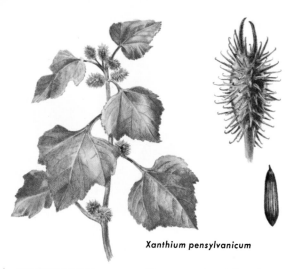

Xanthium pensylvanicum

COCKLEBURS are readily recognized by their prickly bur-like fruits that contain two seeds. The burs help spread the plants by clinging to clothing and to animal fur. There are numerous species growing in North America, all somewhat similar in general appearance. Cockleburs are common throughout the U.S. along roadsides, in fields, and near reservoirs. On river floodplains, they often develop solid stands covering many acres. In the 2-leaf seedling stage, they are extremely poisonous to livestock. The plants can be readily controlled by using herbicides.

Three Cocklebur species are widespread: *Xanthium pensylvanicum*, *X. italicum*, and *X. chinense* (a misnomer; actually from Vera Cruz, Mexico). Spiny Cocklebur, *X. spinosum*, is also widespread but less common.

Rudbeckia serotina

BLACKEYED-SUSAN is generally classed as a wild-flower, but because it aggressively invades pastures and hayfields, sometimes covering them with beautiful blossoms, it is also a weed. The large orange-yellow flowers with dark centers are borne singly on stalks or branches 1 to 3 feet tall. The popular Gloriosa Daisies were developed recently from these attractive plants. Blackeyed-susans are spread by their oblong 4-angled seeds which do not have a parachute, or pappus, to aid their distribution. As it invaded the East, this plant has given rise to many freakish forms.

Blackeyed-Susan, *Rudbeckia serotina* is the only widespread and weedy species of about 40 members in this genus in N.A. Known also as Yellow Daisy, this common and familiar plant is also the state flower of Maryland.

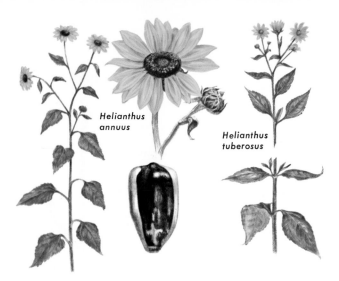

Helianthus annuus

Helianthus tuberosus

SUNFLOWERS, the state flower of Kansas, are present in every part of the U.S. but are most abundant in the plains and prairies. The genus includes about 100 species, mostly North American, annuals and perennials. The lower leaves are opposite, the upper ones alternate. Common weedy species generally grow in pastures, abandoned fields, roadsides, or similar open places. The giant-flowered garden form that turns its head toward the sun is a cultivated variety of a widespread native species. Sunflowers produce flattish 4-angled seeds, a valuable food for birds.

The Common Sunflower is *Helianthus annuus.* Prairie Sunflower, *H. petiolaris,* is a widespread western species. Blueweed, *H. ciliaris,* is a perennial. Jerusalem Artichoke, with edible tubers, is *H. tuberosus.*

134

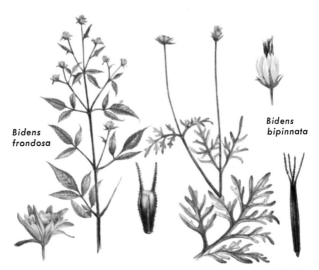

Bidens frondosa

Bidens bipinnata

PITCHFORKS, also known as Spanish Needles, Stick-tight, Beggarlice, and Bur-Marigolds, have distinctive seeds armed with 2 to 4 barb-bearing prongs that hold tenaciously to clothing and fur, thus spreading them. These annual or perennial herbs are generally 2-4 feet tall, with the opposite leaves toothed or incised in some species, pinnately compound in others. The flowers are mostly yellow, white in a few. About 25 species occur in the U.S., mainly in the East; a few are widespread. Most prefer damp habitats, but several are weeds in comparatively dry soil.

Common weedy species in relatively dry soils include *Bidens frondosa*, *B. aristosa*, *B. comosa*, and *B. bipinnata*. Those common to moist soils include *B. cernua* and *B. laevis*. *B. beckii* is an uncommon aquatic.

135

GALINSOGA, though common in gardens and else-where, has no recognized common name. Its scientific name honors a Spanish botanist. This rather unattractive annual herb of the daisy family (Compositae) usually grows a few inches to a foot or more tall. It has opposite, thin, ovate leaves, and the white-and-yellow flowerheads are small. Of half a dozen species in the genus which is native to Tropical America, at least four occur in various parts of the U.S. They are partial to rich moist soil. Galinsoga is rarely abundant enough to be a serious pest.

Common species in the U.S. are *Galinsoga ciliata* and *G. parviflora*. *G. caracasana* and *G. bicolorata* occur more sparingly. They all spread by seeds (achenes).

TARWEEDS include heavy-scented sticky species of the two closely related western genera: *Madia* and *Hemizonia*. Both are especially abundant in California. The features distinguishing the two genera are minor and technical. Often the distinct, "foxy" odor of Tarweed can be detected in passing pastures or fields where the plants are growing in abundance. Their leaves are usually linear, the petals (rays) three-notched and white or yellow. The dark (often black) seeds of Tarweed are arched and triangular in shape. They are eaten by both songbirds and gamebirds.

Common Tarweed, *Madia elegans*, Chilean Tarweed, *M. sativa*, Hayfield Tarweed, *Hemizonia congesta*, Virgate Tarweed, *H. virgata*, and Coast Tarweed, *H. corymbosa*, are the most important species.

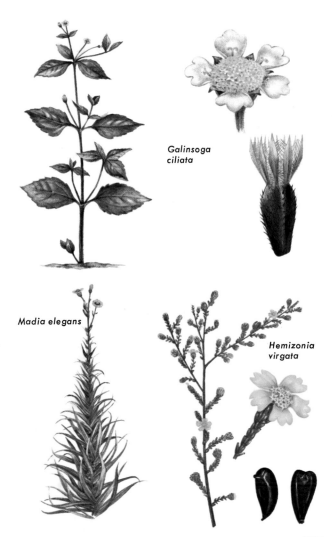

*Galinsoga
ciliata*

Madia elegans

*Hemizonia
virgata*

137

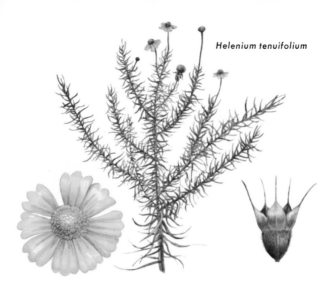

Helenium tenuifolium

BITTERWEED, also called Sneezeweed, is one of the most prevalent weeds of the South. It is particularly common in pastures, along roadsides, railroads, and in idle lots. A principal objection to this graceful, approximately one-foot annual is its consumption by cows in the spring, for it makes their milk taste bitter. Cattle are suspected of helping to spread the plants by distributing their seeds, which are also windblown because of their parachute-like appendages, consisting of hairs and scales. The plants have lacy leaves and attractive bright yellow flowers.

Bitterweed, *Helenium tenuifolium,* is the only serious weed of approximately 25 North American species in this genus. Other common species are *H. nudiflorum* and *H. autumnale.*

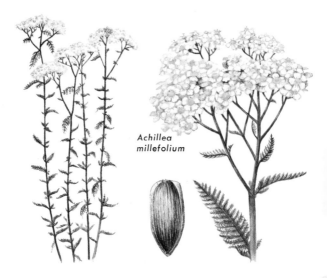

Achillea millefolium

YARROW, also called Milfoil, Thousandleaf, and Woundwort, is a perennial native. About 100 other species occur mainly in Europe and Asia. This is one of the most familiar weeds near old buildings, in pastures, and along roadsides. At one time it was a popular medicinal herb used for stopping the flow of blood. A tea was brewed from the leaves as a cure for colds. The plants are generally 1 to 2 feet tall, with alternate, pungent, finely dissected, fuzzy foliage and a flat-topped cluster of tiny white flowers. Milk and milk products are tainted when cattle eat plant.

Common Yarrow, *Achillea millefolium,* is also native to Eurasia. A pink-flowered form is grown as an ornamental. Western Yarrow, *A. lanulosa,* is a less important weed.

MAYWEED belongs in an Old World genus of about 100 species. This familiar weed, which occurs in all parts of the U.S., is partial to rich moist soil or barnyards, fallow fields, and roadsides. The plants are usually a foot or less tall, with pungent finely cut leaves and yellow-centered white flowers that bloom in early summer. Also known as Dog Fennel or Chamomile, Mayweed commonly occurs in abundance locally, making masses of white flowers. The roundish yellow centers (disks) persist long after the white petals (rays) have withered and dropped.

Mayweed, *Anthemis cotula*, is the most common species. Field Chamomile, *A. arvensis*, is locally plentiful; its leaves are not ill-scented. Neither species is ordinarily a serious weed.

GROUNDSEL, or Ragwort, belongs to a large genus of about 1,200 species, yet few people know these common weeds. Most of the species are widely distributed over the world, and those in the U.S. include both natives and exotics, perennials and annuals. They are serious pest plants because they are poisonous to grazing livestock. The plants usually grow 1 to 2 feet tall, with lobed or toothed leaves and attractive bright yellow flowers. Their seeds have soft-hairy white parachutes, easily carried by the wind. Groundsels are common in pastures and fallow fields.

Common Groundsel is *Senecio vulgaris*. Other important species are Riddell Ragwort, *S. riddellii*, Small's Groundsel, *S. smallii*, Platte Groundsel, *S. plattensis*, and Stinking Willie, *S. jacobaea*.

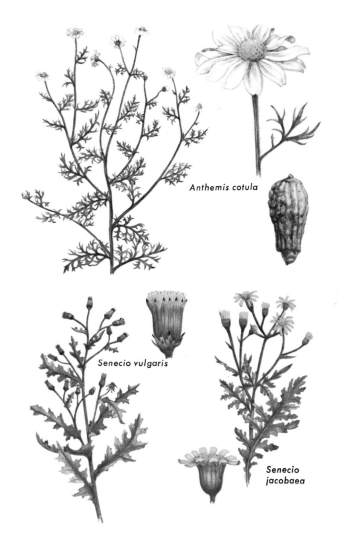

Anthemis cotula

Senecio vulgaris

*Senecio
jacobaea*

Chrysanthemum leucanthemum

FIELD DAISY, or Oxeye Daisy, is both a wildflower and a weed. In many places, especially in the Northeast, pastures and hay meadows are ruined by the abundance of this attractive plant. Cattle do not like its bitter taste.

Field Daisy, a perennial immigrant from Europe, belongs to the same genus as the Shasta Daisy and Chrysanthemum, along with about 159 other species. The plants are usually about 2 feet tall with solitary, showy, white-rayed flowers at the ends of branches. The seeds are oblong, with 8 to 10 lengthwise ribs.

Field Daisy is *Chrysanthemum leucanthemum*. Though cattle do **not** relish the plants, they sometimes do eat them, causing their milk to taste bitter. Mowing helps control the plant.

Arctium minus

BURDOCK, in the composite family, represents a genus of a half-dozen species from the Old World. Two species are widespread but only locally abundant through the northern states. They are rarely serious weeds. These coarse, rank-smelling biennials are partial to rich soil near old farm buildings and along fence rows. Varying from 2 to 5 feet tall, the plants have large oval or heart-shaped leaves as much as a foot long and purplish flower heads covered by long recurved bristles, constituting the "bur." The oblong seeds are flattish and have a lengthwise ridge.

Common Burdock is *Arctium minus*. Great Burdock, *A. lappa*, a weed, is cultivated by Japanese, who use roots of young plants as a vegetable. Livestock rarely eat toxic foliage.

143

Cirsium arvense

THISTLES are one of the most familiar types of weeds because of the spines on their leaves and flower heads. There are about 200 species in the world, mainly in the Northern Hemisphere. Numerous species, both native and exotic, occur in various parts of the U.S. and may be abundant in pastures and along roadsides. Neither sheep nor goats will feed on their spiny stems or leaves.

Thistle plants vary from 2 to as much as 6 feet tall. They have alternate lobed leaves edged with numerous spines and large attractive flower heads of purple, rose, pink, or occasionally yellow or white. Many of the species are aggressive serious weeds, but probably the worst of all is Canada Thistle (actually of European

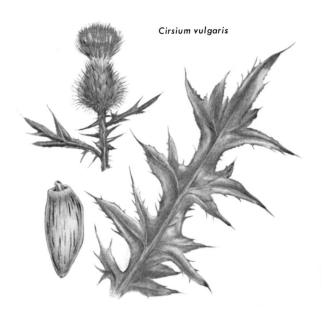

Cirsium vulgaris

origin). Almost any part of the Canada Thistle's root-stock can start a new plant. It is for this reason that the weeds are found in ever-spreading patches, and are so difficult to eradicate. Thistle seeds are carried through the air by elaborate bristly plumes, or parachutes. They are a favorite food of Goldfinches.

Canada Thistle is *Cirsium arvense*. Among the other common species, though not as troublesome or as widespread, are: Bull Thistle, C. *vulgaris*, Western Thistle, C. *occidentale*, Wavyleaf Thistle, C. *undulatum*, Indian Thistle, C. *edule*, and Tall Thistle, C. *altissimum*.

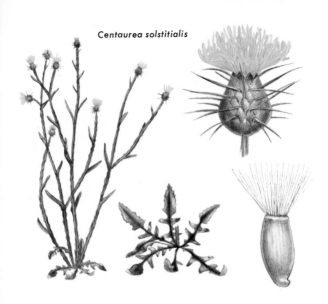

Centaurea solstitialis

STAR-THISTLE as used here refers to two groups of plants: Star-Thistles proper and the Knapweeds. Both belong to the genus *Centaurea* of about 500 Old World species. Included are numerous pernicious pests, both perennials and annuals. Many of the annuals are spiny or thorny. The group has representative species in all parts of the U.S., but they are most abundant on the Pacific Coast, particularly in California.

Star-Thistles of several species, confined largely to the Pacific Coast, are thorny. They are found in pastures, cultivated fields, orchards, and waste places. Their flowers vary from yellow to purple.

Russian Knapweed is a serious weed in many western states, commonly occurring in croplands as well as in fallow fields. Like Bindweed and Canada Thistle,

146

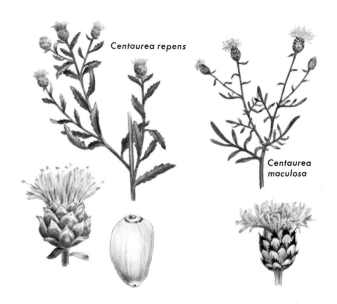

Centaurea repens

Centaurea maculosa

this plant is difficult to control because of its horizontal rootstocks that produce new plants every here and there. The purplish, thistle-like flower heads lack spines. The perennial Brown Knapweed and the biennial Spotted Knapweed are other members of this group.

Yellow Star-Thistle, *Centaurea solstitialis*, is a particularly common species on the Pacific Coast. Others are Napa Star-Thistle, *C. melitensis*, Purple Star-Thistle, *C. calcitrapa*, and Iberian Star-Thistle, *C. iberica*.

Russian Knapweed is *C. repens*, Brown Knapweed, *C. jacea*, and Spotted Knapweed, *C. maculosa*.

Silybum marianum

MILK-THISTLE is so named because of milky-white veins in its leaves. It belongs to the genus *Silybum*, separate from the true or ordinary Thistles, *Cirsium*, but is thistle-like in many ways. This Mediterranean biennial occurs sparingly in many states but is common on the Pacific Coast, particularly in California where it is abundant in pastures, cultivated fields, and along roadsides or ditch banks. The plant stands 2 to 6 feet tall, has basal leaves as much as two and a half feet long and about half as wide, and is topped by large spiny purple flower heads.

Milk-Thistle, *Silybum marianum*, is spread by large smooth seeds that are carried in the wind. Each seed bears an elaborate parachute of bristly hairs. Milk-Thistle is sometimes grown as an ornamental.

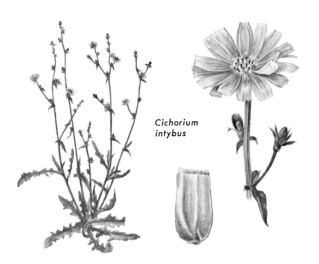

*Cichorium
intybus*

CHICORY, also called Blue Sailors, Blue Daisy, and Coffeeweed, is a Eurasian perennial that is clearly a weed and sometimes a serious pest. When its large blue flowers (on nearly leafless branches) are open in the morning, it also qualifies as an attractive wildflower. Chicory is locally common to abundant across the U.S., mainly along roadsides, in vacant lots, pastures, and waste places. The plant has a deep taproot, milky juice, and a basal rosette of lobed toothed leaves. Its seeds are angular and flat-topped. Used locally as an additive to coffee.

Chicory, *Cichorium intybus,* has a cultivated relative, Endive, C. *endivia,* that occurs in a cut-leafed form or with long lettuce-like leaves. Endive is used like lettuce in salads. It has a strong, somewhat bitter taste.

149

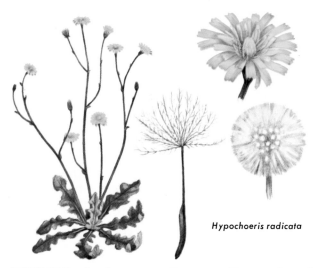

Hypochoeris radicata

CAT'S-EAR, also known as Gosmore, Flatweed, and Coast Dandelion, occurs sparingly across the northern U.S. but is abundant in some parts of California. This Old World perennial is partial to lawns, pastures, meadows, and waste places. It usually grows 1 to 2 feet tall and, like its close relative, Dandelion, has a flattish rosette of hairy leaves and the inclination to be a lawn pest. The yellow flower heads, about one inch across, are borne on long, branched stalks. Mowing off flowering tops does not control the plants but does help prevent seed distribution.

Cat's-ear, *Hypochoeris radicata*, is the only weed of this genus in the U.S. Its 10-ribbed seeds are carried by the wind, floating by a hairy plume, or parachute, attached to the seed by a long slender beak.

SALSIFY goes by numerous other names such as Oysterplant, Goats-Beard, Noonflower, and Jerusalem Star. The tall (usually 2-4 ft.) gray-leaved plants with milky sap were introduced from Europe for their large, edible roots. Now naturalized and widespread in North America, it thrives in backyards, vacant lots, and along roadsides. Its showy purple or yellow flowers open in the morning and close by noon, hence the name "Noonflower." Like many members of the daisy family (Compositae), the seeds are equipped with parachutes. Ordinarily it is not a serious pest anywhere.

The purple-flowered *Tragopogon porrifolius* is particularly abundant and widespread, but the two yellow-flowered species, Meadow Salsify, *T. pratensis*, and the similar *T. major* are also common.

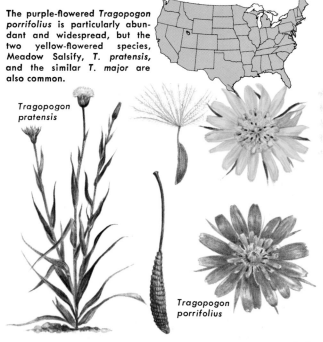

Tragopogon pratensis

Tragopogon porrifolius

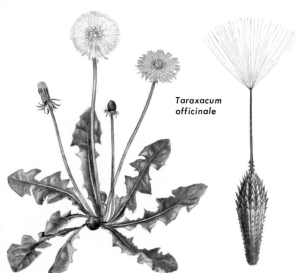

Taraxacum officinale

DANDELION is one of our most familiar weeds. Most people who live in the temperate parts of the U.S. see this weed almost every day. It abounds in meadows, yards, along roadsides, and elsewhere but is especially conspicuous in lawns. Because of its deep taproot, the plant is difficult to destroy, and the balls of parachute-borne seeds spread it to new locations. Dandelion's fragrant, attractive flowers are used to make wine; the milky juice of the roots has medicinal values; and the leaves are often used for greens. Bees use the nectar in making honey, and songbirds like the seeds.

Common Dandelion is *Taraxacum officinale*. Red-seeded Dandelion, *T. erythrospermum*, is widespread but usually less abundant. Species range into Arctic and southernmost parts of Southern Hemisphere.

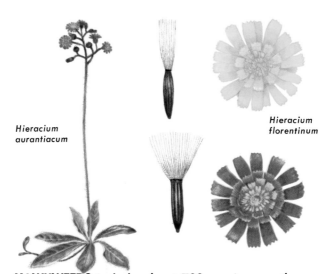

Hieracium aurantiacum

Hieracium florentinum

HAWKWEEDS include about 700 species, mostly native to Europe and South America; a few are sometimes grown as ornamentals. The native North American species are harmless woodland plants. A few European species of Hawkweeds have become abundant in the Northeast and are troublesome weeds in pastures, meadows, and occasionally in lawns. Typically, these members of the daisy family (Compositae) are perennials that have a branched flower stalk, from 6 inches to 2 feet tall, arising from center of a rosette of hairy leaves. Mowing reduces flowering and retards spread.

Orange Hawkweed, also called Devil's Paintbrush, *Hieracium aurantiacum,* is a beautiful but pernicious pest. Yellow-flowered relatives such as King Devil Hawkweed, *H. florentinum,* are less attractive.

153

Sonchus oleraceus

SOW THISTLES total about 45 Eurasian species, half a dozen of which occur in the U.S. Of these, three are particularly widespread. They are so common in city lots and along sidewalks, as well as on farms, that practically everyone sees these plants daily. They get little attention, however, simply because they are unattractive weeds.

Sow Thistles usually grow 1 to 3 feet tall. They are coarse leafy-stemmed weeds with toothed or prickly-lobed alternate leaves and pale yellow flowers. One species, Perennial Sow Thistle, has bright yellow flowers. Another feature that makes them unattractive is their bad-tasting milky juice.

The seeds of Sow Thistles are ribbed lengthwise and have a parachute composed of numerous soft, fine

154

Sonchus asper

Sonchus arvensis

cottony-white hairs. Birds are fond of the seeds. Most species spread mainly by the wind-blown seed, but Perennial Sow Thistle also spreads by its lateral roots, which run horizontally near the soil surface, break easily, and bud into new plants. This species is a noxious weed which is generally controlled by chemicals in fields or by regular cultivation in gardens.

Common Sow Thistle, *Sonchus oleraceus*, Prickly Sow Thistle, *S. asper*, and Perennial Sow Thistle, *S. arvensis*, are the most common and important species. Less common are *S. uliginosus* and *S. tenerrimus*. To prevent spread of Sow Thistles, it is important that the plants be killed or mowed before seeds are set.

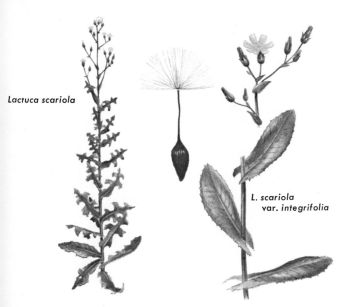

Lactuca scariola

L. scariola
var. *integrifolia*

WILD LETTUCE is a genus of about 100 species, only a few of them native to the Americas. One native American species, the purple-flowered Blue Wild Lettuce, is a perennial. All the other species in the Americas are annuals, and most of them are from Europe.

The most widespread and abundant species is Prickly Wild Lettuce. It is also known as Compass Plant because its bluish-green leaves are twisted slightly at their clasping base so that they mainly face east and west, their edges pointing approximately north and south. The plant is usually 1 to 5 feet tall with small pale-yellow flowerheads. One form has unlobed (entire), toothed leaves, the other has lobes that resemble broad crossbars across the leaf. Both contain milky juice and have earlike lobes at leaf base.

Lactuca pulchella

Lactuca saligna

Other species usually vary from 1 to 6 feet tall. They are locally common in cultivated or fallow fields, vacant lots, and waste places. Seeds of all species are ridged lengthwise and narrowed into a beak at the parachute attachment.

Prickly Wild Lettuce is *Lactuca scariola*. Its unlobed-leaf form is *L. scariola* var *integrifolia*. Blue Wild Lettuce is *L. pulchella*. Other species in the U.S. include Canada Wild Lettuce, *L. canadensis*, Wild Willow-leaved Lettuce, *L. saligna*, *L. muralis*, and *L. biennis*. Garden Lettuce, *L. sativa* (and varieties), is a cultivated species of this genus.

The books and publications listed below are only a few of the many that are useful in identifying weeds. The most important for use locally are probably the various pamphlets and bulletins issued by state agricultural departments. These usually provide specific control measures for destroying particular weeds. If you have an unusually plaguing weed problem, get assistance from your county agent or state agricultural representatives.

USEFUL WEED PUBLICATIONS

Common Weeds of the United States, prepared by USDA, Dover Publications, New York, 1971.

Weeds, W. C. Muenscher, Comstock Publishing Associates, Cornell University Press, Ithaca, 2nd ed., 1980.

All About Weeds, E. R. Spencer, Dover Publications, New York, 1974.

Weeds of the North Central States, K. P. Buchholtz, B. H. Grigsby, O. C. Lee, and others, University of Illinois Agri. Exp. Station, Champaign, 1954.

Weeds of the Pacific Northwest, Helen M. Gilkey, Oregon State College, Corvallis, 1957.

Weed Identification and Control, Duane Isely, Iowa State University Press, Iowa City, 1960.

Modern Weed Control, Alden S. Crafts, University of California Press, Berkeley, 1975.

Weed Control As a Science, Glenn C. Klingman, John Wiley & Sons, New York, 1961.

Weeds of Lawns and Gardens, John M. Fogg Jr., University of Pennsylvania Press, Philadelphia, 1945.

Weeds: Guardians of the Soil, John A. Cocannouer, Devin-Adair, New York, 1950.

Selected Weeds of the United States, prepared by USDA, Government Printing Office, Washington, DC.

MEASURING SCALE (IN MILLIMETERS AND CENTIMETERS)

INDEX

To Principal Common Names And Latin Generic Names

MEASURING SCALE (IN INCHES)

a Golden Guide® FROM ST. MARTIN'S PRESS

Enjoy and Learn!
Expert Knowledge!
Easy-to-Read!

This handy identification guide to the plants that cause billions of dollars annually in crop loss and control measures includes information on:

- **THE HARM THAT WEEDS CAUSE**
- **BENEFITS FROM WEEDS**
- **MAJOR WEED HABITATS**

Accurate full-color illustrations and descriptive text identify the principal weeds and weed groups that invade lawns, gardens, fields, and roadsides. Range maps show distribution within the United States.

A Classic Guide
...for All Ages

$6.95/$10.95 Can.

Cover design by Michael Storrings
Front cover photograph by Earth Scenes © John Gerlach

www.stmartins.com

175 FIFTH AVENUE, NEW YORK, N.Y. 10010
DISTRIBUTED IN CANADA BY H. B. FENN AND COMPANY, LTD.
PRINTED IN CHINA

ISBN 1-58238-160-7

50695

EAN

9 781582 381602